S0-BHR-803

Ecotourism, NGOs and Development

This book takes a critical look at the role of ecotourism in bringing about sustainable development in the developing world. Ecotourism is often advocated as a sustainable option as it combines development with an emphasis on preserving wildlife and cultures. However, as argued in this book, it also ties the development prospects for rural communities to a 'nature first' outlook that severely limits the prospects for substantial economic development.

Ecotourism has been initiated by a range of non-governmental organisations as exemplary sustainable development in the rural developing world. This book looks at the way these NGOs advocate ecotourism, and identifies key features of this advocacy. These features – the emphasis on local community participation and on the role of local tradition, the assumption of environmental fragility and the emphasis on preserving natural capital, and the overarching assumption that development should integrate conservation and development on a local level – are critically evaluated. It is argued that ecotourism's popularity as a development option devalues human development by tying the latter to an externally imposed conservation priority.

Many authors have written critically about the record of ecotourism in successfully involving communities in development, or in conserving biodiversity. However, the general aim of this development strategy – to link the well-being of rural communities with conservation (integrated conservation and development) – is generally taken to be a normative goal. By contrast, this book questions the rationale behind ecotourism integrated conservation and development projects (ICDPs), and argues that it reflects a diminished view of the potential for substantial development and liberation from poverty.

Jim Butcher lectures at Canterbury Christ Church University in Kent. His previous book, *The Moralisation of Tourism* (Routledge 2003), comprised a defence of mass tourism in the face of its many detractors.

Contemporary Geographies of Leisure, Tourism and Mobility

Series Editor: C. Michael Hall

Professor at the Department of Tourism, University of Otago, New Zealand

The aim of this series is to explore and communicate the intersections and relationships between leisure, tourism and human mobility within the social sciences.

It will incorporate both traditional and new perspectives on leisure and tourism from contemporary geography, e.g. notions of identity, representation and culture, while also providing for perspectives from cognate areas such as anthropology, cultural studies, gastronomy and food studies, marketing, policy studies and political economy, regional and urban planning, and sociology, within the development of an integrated field of leisure and tourism studies.

Also, increasingly, tourism and leisure are regarded as steps in a continuum of human mobility. Inclusion of mobility in the series offers the prospect to examine the relationship between tourism and migration, the sojourner, educational travel, and second home and retirement travel phenomena.

The series comprises two strands:

Contemporary Geographies of Leisure, Tourism and Mobility aims to address the needs of students and academics, and the titles will be published in hardback and paperback. Titles include:

The Moralisation of Tourism
Sun, sand . . . and saving the world?
Jim Butcher

The Ethics of Tourism Development
Mick Smith and Rosaleen Duffy

Tourism in the Caribbean
Trends, development, prospects
Edited by David Timothy Duval

Qualitative Research in Tourism
Ontologies, epistemologies and methodologies
Edited by Jenny Phillimore and Lisa Goodson

The Media and the Tourist Imagination
Converging cultures
Edited by David Crouch, Rhona Jackson and Felix Thompson

Tourism and Global Environmental Change
Ecological, social, economic and political interrelationships
Edited by Stefan Gössling and C. Michael Hall

Routledge Studies in Contemporary Geographies of Leisure, Tourism and Mobility is a forum for innovative new research intended for research students and academics, and the titles will be available in hardback only. Titles include:

1 **Living with Tourism**
Negotiating identities in a Turkish village
Hazel Tucker

2 **Tourism, Diasporas and Space**
Tim Coles and Dallen J. Timothy

3 **Tourism and Postcolonialism**
Contested discourses, identities and representations
C. Michael Hall and Hazel Tucker

4 **Tourism, Religion and Spiritual Journeys**
Dallen J. Timothy and Daniel H. Olsen

5 **China's Outbound Tourism**
Wolfgang Georg Arlt

6 **Tourism, Power and Space**
Andrew Church and Tim Coles

7 **Tourism, Ethnic Diversity and the City**
Jan Rath

8 **Ecotourism, NGO's and Development**
A critical analysis
Jim Butcher

Ecotourism, NGOs and Development

A critical analysis

Jim Butcher

LONDON AND NEW YORK

First published 2007
by Routledge
2 Park Square, Milton Park, Abingdon, Oxon, OX14 4RN

Simultaneously published in the USA and Canada
by Routledge
270 Madison Ave, New York, NY 10016

Routledge is an imprint of the Taylor & Francis Group, an informa business

Typeset in Times New Roman by
Florence Production Ltd, Stoodleigh, Devon
Printed and bound in Great Britain by
Biddles Ltd, King's Lynn

British Library Cataloguing in Publication Data
A catalogue record for this book is available
from the British Library

Library of Congress Cataloging in Publication Data
A catalog record for this book has been requested

ISBN10: 0–415–39367–1 (hbk)
ISBN10: 0–203–96207–9 (ebk)

ISBN13: 978–0–415–39367–6 (hbk)
ISBN13: 978–0–203–96207–7 (ebk)

Contents

Preface

The year 2002 was significant in the rise of ecotourism from trendy market niche to green development strategy. This year was designated as the International Year of Ecotourism (IYE) by the United Nations (UN), an event that marked out the arrival of ecotourism as exemplary sustainable development in the rural developing world, a view now endorsed by the highest level of global governance.

The event had many vocal critics. The fear that ecotourism would be equally as destructive as the mainstream tourism industry was a prominent theme. Anita Pleumaron of the Third World Network's Tourism Monitoring Center feared the event may 'encourage all holidaymakers to become eco-tourists, resulting in hordes of travellers invading villages and protected areas', a scenario that 'could not be called "sustainable" and would create more undesirable impacts to add to the vast problems already found in existing organised tourism' (Pleumaron undated). Nina Rao, southern co-chair of the NGO Tourism Caucus at the UN Commission for Sustainable Development concurred: 'I really think this is going to be worse than the launch of package tours to the Third World' (cited in Pleumaron undated). This opposition fitted the common pattern of criticism of ecotourism – that it is, or threatens to be, a Trojan horse piercing the defences of erstwhile pristine environments and indigenous cultures. Further, fears that ecotourism may 'open the doors to more forest destruction' (Pleumaron undated), 'destroy more biodiversity and harm more local communities' (Ling, cited in Pleumaron undated) or promote 'opportunities for a whole range of investors to gain access to remote rural forest, coastal and marine areas' (ibid.) were characteristic of the opposition to the UN's initiative.

Conspicuous by its absence was any voice pointing out that ecotourism offers only the most meagre prospects for economic development for impoverished societies. Neither did anyone point out that the problems created by ecotourism can be viewed as a product of too little and too partial a development, or by poverty itself, rather than too much rapacious development in the fashion suggested by the critics cited above. This characteristic of the debate is not surprising in that thoroughgoing, transformative economic development – development that has the capacity to change the way societies relate to the natural world through modern technology – has been ruled out of order through a particular, ecocentric interpretation of 'sustainable development', one that is routinely characterised by an insistence on a localised harmony, or 'symbiosis', between human needs and the environment.

Notably, the heat generated by the fraught debate over the merits of ecotourism focuses around operational features, rather than questions of principle. Some argue the community needs more say in the development of ecotourism projects, others argue that environmental considerations may lack priority, and others still remain suspicious of a global industry such as tourism even when wrapped in the green aura of ecotourism. Yet the ideology behind ecotourism – that sustainable development involves a mutually reinforcing, or symbiotic, localised relationship between people and environment – is shared across all protagonists in the debate. It is this ideology that this study takes issue with. It is an ideology that, while rhetorically people centred, stressing 'empowerment' and 'community', involves tying the development prospects for these same people to severe localised natural limits . . . in the name of sustainable development.

The ideas expressed are contrary to received wisdom in some circles. It is hoped that they will contribute to a greater emphasis on the possibilities for development well beyond the localised 'natural' limits that underpin the advocacy of ecotourism as sustainable development . . . limits that are better regarded as social, and hence impermanent.

The author wishes to thank Pete Smith, Professor Kevin Hannam, Barbara Smith, Joanna Williams, the interviewees featured in the book and others who have discussed the ideas in the course of their development.

Abbreviations

ANWR	Arctic National Wildlife Refuge
APPA	Appreciative Participatory Planning and Action
Campfire	Communal Areas Management Programme for Indigenous Resources
CBNRM	community-based natural resource management
CBT	community-based ecotourism
CCA	Conservation Corporation of Africa
CELB	Center for Environmental Leadership in Business
CI	Conservation International
CITES	Convention on International Trade in Endangered Species
CSD	Commission for Sustainable Development
DfID	Department for International Development
DHF	Dag Hammarskjöld Foundation
DNPWM	Department of National Parks and Wildlife Management
ECTWT	Ecumenical Coalition on Third World Tourism
GATS	General Agreement on Trade in Services
GEF	Global Environmental Facility
GIS	geographic information system
ICDPs	ecotourism integrated conservation and development projects
IIED	International Institute for Environment and Development
IMF	International Monetary Fund
IRDNC	Rural Development for Nature Conservation
IUCN	International Union for the Conservation of Nature
IYE	International Year of Ecotourism
LIFE	Living in a Finite Environment
MAB	Man and the Biosphere
MLGRUD	Ministry of Local Government, Rural and Urban Development
NGO	non-governmental organisation
NPA	New Policy Agenda
NSM	new social movement
PRA	Participatory Rural Appraisal
RDCs	Rural District Councils
SAPs	structural adjustment programmes

SIDS	Small Island Developing States
SNV	Stichting Nederlandse Verijwilligers
TEN	Tourism European Network
TIES	The International Ecotourism Society
UN	United Nations
UNDP	United Nations Development Programme
UNEP	UN Environment Programme
USAID	United States Agency for International Development
VSO	Voluntary Service Overseas
WCED	World Commission on Environment and Development
WCS	World Conservation Strategy
WSSD	World Summit on Sustainable Development
WTO	World Tourism Organisation
WTTC	World Travel and Tourism Council
WWF	World Wide Fund for Nature

1 The study and its premises

Introduction

In the 1990s integrated conservation and development projects (ICDPs) linked to tourism emerged as a tool for combining development with conservation, particularly in rural parts of the developing world (Ghimire and Pimbert 1997). A significant number of non-governmental organisations (NGOs) of various kinds – some whose principal interest is in conservation, others with aims focusing on development and community well-being – adopted this innovation, proposing it to constitute sustainable development.

Subsequently, debates about the merits of ecotourism have attracted much heat, but less light. Notably, while there has been much fraught debate on the ability of such projects to deliver on their stated aims, these aims – limited development for the community based on their involvement in the conservation of surrounding natural resources – have been taken to be normatively good, this underpinned by their association with sustainable development.

This study critiques ecotourism as a tool for integrated conservation and development in the developing world, principally from the standpoint of human development. In order to do this, the study adopts a dual focus on the practitioners (NGOs who have pioneered this approach in their work since the 1980s) and on academic and related literature that implicitly or explicitly advocates ecotourism as having the potential to constitute exemplary sustainable development. Taken together they comprise an influential strand of thinking on development and conservation, and an influential discourse on rural development in the developing world.

In this introductory chapter, the origin, terms of reference and premises of the study are established, and the arguments to be developed are set out.

The origins of the study

Tourism is one industry that has been subject to much scrutiny for its perceived lack of sustainability, and 'sustainable tourism' has become widely advocated for its potential to be benign, and even positive (Hall 1998: 13–24).

Ecotourism ICDPs emerged around the beginning of the 1990s as one way of achieving what its advocates regard as sustainable tourism development. Aid funded projects involving ecotourism are often referred to as ICDPs, as they seek to combine these two goals – conservation and development – and overcome the tension often considered to exist between them. NGOs of both a conservation and community well-being bent respectively have taken up ecotourism to this end.

It is somewhat ironic that tourism should be seen as having the potential to combine development with conservation, given the popular caricature of mass tourism as the destroyer of once quaint fishing villages and proud cultures (Butcher 2003a: 24–5 and 38). Yet the logic, on the face of it, is straightforward. Prominent advocate of ecotourism, Harold Goodwin, makes the case for ecotourism thus in a paper entitled 'Tourism and Natural Heritage: a Symbiotic Relationship?' (Goodwin 2000). Goodwin argues against the view that tourism, like any other industry, is necessarily in conflict with the natural environment. Rather, he argues, it has a special role to play in development. Ecotourism, depending as it does on a desire to experience areas of perceived natural beauty and distinction, can provide funds to manage and maintain conservation areas. Revenues can also serve the function of encouraging local communities to cooperate in conservation. This outlook may be especially apt given the growth of the ecotourism market in the developed world, where a desire to be 'ethical' through experiencing a closer relationship with the natural world appears to be a significant and growing push factor on demand (Poon 1993; WTO 2003).

However, though this formulation has its own clear internal logic, that logic rests on certain assumptions that themselves may be questioned. Indeed, any formulation of sustainable development will inevitably reside on prior conceptions of the natural world, development, and the relationship between the two, and these are contested (Potter *et al.* 1999; Urry and MacNaghten 1998; Pepper 1996: 74–5; Hughes 1995; Redclift 1990). As Mowforth and Munt have it, sustainability is not value neutral, but 'a concept charged with power' (1998: 25), and those invoking it are usually asserting a distinctive position rather than a universally agreed viewpoint (ibid.). The motivation for the study was to examine the prior assumptions underpinning the advocacy of ecotourism ICDPs as sustainable development in the rural developing world, and to make these assumptions explicit in order to critically analyse them. In so doing, it is hoped to broaden out what has thus far been generally a technical debate about how to organise and run projects, and also to place ecotourism in the context of trends in thinking on development.

The study in brief

Formally stated, the aim of the study is to examine the advocacy of ecotourism as exemplary sustainable development in the rural developing world, and offer a critical analysis of the assumptions underlying this advocacy. The objectives undertaken to meet this aim are: to identify a sample of case study NGOs, selected purposively, on the bases of their experience in this area and their capacity to broadly reflect the wider population of pertinent NGOs; to establish the ideological context for ecotourism ICDPs by situating them in developments in post-Second World War conservation and development thinking; to critically analyse the assumptions which are at the heart of the advocacy of ecotourism ICDPs as sustainable development in the developing world (both in the case studies and in the general literature); and to consider the implications of the strategy, with regard to how development is conceived of.

Following this introductory chapter, Chapter 2, titled 'Ecotourism in development perspective' examines influential literature on conservation and development respectively. The convergence of important strands of each around *neopopulist* themes related to local community participation is characteristic of the advocacy of ecotourism. Hence, as an innovation in development practice ecotourism is exemplary of wider trends in development thinking, and therefore the study can be read as a comment on this bigger picture too. The chapter also positions this study in relation to a selection of pertinent ecotourism specific literature. Overall, it attempts to provide an ideological context for the contemporary advocacy of ecotourism, placing it in the trajectory of wider thinking on development and conservation.

Chapter 3, titled 'Pioneers of ecotourism: different aims, shared perspective', attempts to summarise the aims of the organisations featured in the case studies utilised throughout the book, and also to establish their relationship to the strategy of pursuing sustainable development through ecotourism. It attempts to lead from an outline of the case studies into an analysis of ecotourism ICDPs – or from *who they are* to *what they advocate*. For each case study in turn, the chapter looks at how ecotourism emerged onto their agenda and developed, and makes provisional comments on these agendas. These include establishing what common and divergent themes are evident in the advocacy of ecotourism thus far. Importantly, a symbiosis between conservation and development is clearly identified as the key, common, overarching argument. Hence, the rationale for this symbiosis is set out here.

Chapters 4–6 look in turn at three key assumptions underlying 'symbiosis'. Chapter 4, 'Community participation in the advocacy of ecotourism', focuses on the invocation of local community participation as a central feature, and provides a critical analysis of this. The claims that the neopopulist 'community' emphasis of ecotourism is, if carried through, a progressive alternative to previous modes of development, is criticised, as is the localism that consistently accompanies this emphasis. It is argued that there is a neglect of wider development, beyond the local level, and that the democratic and radical credentials of the community participation agenda mask a profound limiting of the agency of the community. Chapter 5, 'Tradition in the advocacy of ecotourism', identifies and critically considers the emphasis on *tradition* in the discourse. Again, as with the community participation emphasis, the stress put on development on the basis of tradition and traditional knowledge presents ecotourism as involving the agency of the community – it is *their* tradition and hence *their* development. The chapter suggests, in contrast, that the emphasis on tradition reflects a cultural relativism that deprioritises development on any transformative scale (any scale that might undermine tradition) and hence in the name of cultural *difference* neglects the humanist aspiration for material *equality*. Chapter 6, 'Natural capital in the advocacy of ecotourism', identifies the non-consumption of natural capital and the assumption of environmental fragility as important and consistent characteristics of the debate, and analyses the implications of this for economic development. The grand claims made for ecotourism as exemplary sustainable development are often based on its ability to bring revenue (and development) on the basis of the non-use of natural resources. But what kind of development is on offer here . . . and more importantly, what kinds of development are ruled out of court? All the aforementioned analysis is based upon the discourse emanating from the case studies and exemplary academic literature.

The penultimate chapter, Chapter 7, titled 'Symbiosis revisited', attempts to re-present the rationale put forward for ecotourism, introduced in the current chapter and elaborated in Chapter 3, in the light of the critique developed through Chapters 4, 5 and 6. Finally, a brief concluding chapter offers a summary of the study, some general comments and some suggestions as to a provisional research agenda to develop the book's themes.

A note on terminology

A brief note on essential terminology is necessary, as some key terms are used in different senses depending on context. All the terms mentioned below are elaborated upon in greater detail elsewhere in the study.

Sustainable development and sustainable tourism development

There is much debate on what the emphasis of sustainable development should be, and even on its substance. However, the most common definition was that established by the World Commission on Environment and Development (WCED) in its report titled *Our Common Future* in 1987, and popularised at the UN Summit on Environment and Development held in Rio in 1992. This definition states that sustainable development is development that 'meets the needs of the present without compromising the ability of future generations to meet their own needs' (WCED 1987: 43). It sets out inter-generational equity as a central theme of sustainability. However, *Our Common Future* itself was explicit that the meaning of sustainable development remained uncertain and contested (Redclift 1990).

Sustainability has always lacked conceptual clarity, and been interpreted in different ways (Seers 1996 and 1997). Over seventy definitions have been proposed (Steer and Wade-Gery 1993). One author notes that 'people from many diverse fields use the term in different contexts and they have very different concepts, approaches and biases' (Heinen 1994: 23). Indeed, for some the notion contains within it an inherent contradiction, that between development and conservation (Worster 1993; Redclift 1990).

Sustainable development has been described as the 'parental paradigm' of sustainable tourism (Sharpley 2000: 1), and, as such, sustainable tourism is the development of tourism that meets the standard of sustainable development more broadly. Definitions of sustainable tourism are then, unsurprisingly, also numerous (Sharpley 2000; Garrod and Fyall 1998). This is in part due to the way in which sustainability has increasingly moved away from being a narrowly environmental concept, rooted in a perceived scientific understanding of the damage caused by development, towards embodying the relationship between culture and environment (Mowforth and Munt 1998: 43). This change is significant in relation to this study, given the way ecotourism ICDPs explicitly link community well-being and development to the 'sustainable' use of the local natural environment.

A feature of the definitions of both sustainable development and sustainable tourism development is that they tend to be very broad, and there is ample scope for different interpretations. Some of these interpretations give greater emphasis to environmental conservation and others may prioritise economic development. One argument running through this study is that economic development has a low priority when sustainable tourism development is considered with regard to the rural developing world.

However, this study does not offer up a rival definition of sustainable tourism development to add to the plethora that exists. Rather, its purpose is to *problematise* the term in a particular context, that of ecotourism ICDPs in the developing world. It looks at the assumptions in the advocacy of ecotourism as sustainable development, and hence the nature of sustainable tourism development thus constituted, and tries to make these assumptions explicit in order to interrogate them. Hence the study can be regarded as a critique of the way sustainable tourism is interpreted and applied in the rural developing world through ecotourism.

Ecotourism and ecotourism ICDPs

Ecotourism ICDPs are projects that seek to utilise ecotourism as a tool to combine conservation with development opportunities for rural communities, normally in the developing world (Scheyvens 2002: ch. 6). Put simply, ecotourists may pay large amounts of money to visit sites of natural beauty, and the revenue they generate may in turn make the conservation of these sites economically beneficial for the communities within and around them. Ecotourism is especially suited for ICDPs as it involves the *non-consumption* of natural resources (Fennell 2003: 43), as opposed to most forms of development, which by their nature require the transformation or destruction of aspects of the environment. Following this logic, the advocates of ecotourism argue that it can constitute exemplary sustainable development in rural parts of the developing world (Fennell 2003; Scheyvens 2002; Goodwin 2000; Ziffer 1989).

Ecotourism is a sector of the leisure travel market, but one that seems to have purpose beyond the satisfaction of consumer needs. It is often regarded as exemplary 'ethical' tourism, due to the perception that it can combine development and conservation goals, at the same time as providing fascinating opportunities for the tourist (Butcher 2003a: Ch. 1). There is, as with sustainable tourism, some debate over what ecotourism is and is not, and over what it should and should not be (Fennell 2003: Ch. 2; Scheyvens 2002: 69–71). Some argue a more purist, preservationist line in which the environment is paramount, while many more adopt a conservationist approach, allowing for limited change to the environment. Some emphasise the natural environment as the principal attraction (e.g. Fennell 2003: 43), while elsewhere local cultures are prominent alongside the environment (e.g. Wearing and Neil 1999).

However, a typical definition is that of The International Ecotourism Society (TIES), the principal trade body for this market niche:

> Purposeful travel to natural areas to understand the culture and natural history of the environment, taking care not to alter the integrity of the ecosystem, while producing economic opportunities to make the conservation of natural resources beneficial to local people.
>
> (cited in Goeldner *et al.* 1999: 556)

In similar vein, the world's biggest conservation organisation, and one of the five case studies featured in this study, the World Wide Fund for Nature (WWF), defines ecotourism as 'tourism to protect natural areas, as a means of economic gain through natural resource preservation' (ibid.). Both definitions state that ecotourism should address both environmental conservation and the well-being of the local community being visited, and are typical of the definitions widely adopted, including those by the featured NGOs.

Hence 'ecotourism ICDPs' and 'ecotourism' refer to *different sides of the same phenomenon*. 'Ecotourism ICDPs' refer to the supply side – what is going on in the community, at the destination – which is obviously important for development and conservation. However, the projects would not be viable without ecotourists, so 'ecotourism' refers to the demand side – the growing market to travel for leisure to such project areas. Notwithstanding this distinction, the term 'ecotourism' is generally used in this study for simplicity, with the suffix 'ICDPs' added to emphasise its role in development where this adds clarity. As 'ecotourism' is a term that has wider resonance outside of conservation and development speak, it is an appropriate term of reference in an interdisciplinary study such as this.

It should be noted that NGOs and academic publications use a variety of terms for ecotourism, such as 'community based ecotourism', 'community based natural resource management' (CBNRM) and 'ecotourism projects'. Some are incidental, and some, no doubt, seek to mark out a subtle difference from the way other terms are perceived. This is certainly the case, for example, with Tourism Concern, whose *Community Tourism Guide* introduces their favoured term – 'community tourism' – as distinct from some of the 'nature first' connotations of ecotourism (Tourism Concern/Mann 2000). Yet the people-centred approach they set out is shared by other organisations adopting 'ecotourism', 'community based ecotourism' and a host of other alternatives as their preference. This study argues that although they may differ on the detail, all the variations on the ecotourism theme share the same fundamental rationale.

Importantly for this study, ecotourism reflects a model of development that is closely allied to a wider philosophy – it is often referred to as a 'concept' rather than simply another industry niche or green brand. For example, ecotourism pioneer Ziffer, in her early and influential book on the subject, sees ecotourism as a concept that 'ambitiously attempts to describe an activity, set forth a philosophy, and espouse a model of development' (Bottrill and Pearce 1995: 45).

To take Ziffer's definition clause by clause: that ecotourism is an activity is unproblematic – it has been examined as a growing industry niche, as a form of New Tourism (Poon 1993). As an activity, it can be observed and measured, and many researchers have pursued this.

Ecotourism as a philosophy is often referred to or implied in academic publications, NGO advocacy and tour operator brochures. Its philosophical premises are considered in this study – principally it is argued that it embodies a specific notion of human–environment relations, one characterised here as ecocentrism dressed up as a humanism through the rhetoric of 'people', 'community' and 'empowerment'.

Ecotourism as a model of development is also a focus of this book. Were ecotourism simply a Western sensibility feeding into a growing new niche for the tourism industry, then although it would remain an important aspect of contemporary culture to be studied and researched, it would have limited import beyond the culture of the tourists themselves. However, ecotourism has become, or become a part of, a model for development – the ideas comprising the 'philosophy of ecotourism' are influential in rural development, and hence have great importance beyond a rarefied discussion of culture. Specifically, its emphasis on harmony or symbiosis between locally situated human and environmental goals, and the association of this with sustainable development, has strong implications for how development is envisioned.

Non-governmental organisations and civil society

NGOs refer simply to formal organisations that are neither part of the state nor profit maximising commercial companies. There are obviously countless such organisations – everything from the Boy Scouts to human rights organisations, from self-help groups to political parties. However, there is a discernible literature that looks at NGOs as organisations that seek to influence politics from the perspective of civil society, or from outside the direct involvement of government (e.g. Hilhorst 2003; Wallace and Lewis 2000; Princen and Finger 1994). Many such organisations are concerned with the environment, and many others are concerned with issues relating to well-being and development. The list is vast, but it is from within this broad category that ecotourism ICDPs have emerged and been advocated since around 1990.

Civil society is a broader concept. It can be utilised simply to refer to the sum total of associational life outside of the state and of commercial activity (Kumar 1993; Seligman 1992). This abstract term can refer to the involvement in and engagement with important issues by the population, often through formal groups such as NGOs (Hann and Dunn 1996: introduction). As a consequence, civil society is often used almost interchangeably with the term NGOs, and there is usually at least a close implied relationship between the two. A good example of this close relationship, pertinent to this study, is to be found in Scheyvens' *Tourism for Development: Empowering Communities* (2002: 17, 53 and 62).

However, civil society is often also used in a *normative* sense, as an area of human agency relatively free from political dogma and from moneyed interests, and hence able to express a genuine popular subjectivity (Kumar 1993). Civil society in this normative sense is akin to Putnam's (2000) concept of social capital, which refers to people's engagement with each other, with communities, with culture and with society at large through formal and informal bonds. In this sense, civil society has a positive moral association vis-à-vis governments and companies (Potter *et al.* 1999: 177).

This study looks into five pertinent NGO related case studies, as it is from NGOs that the strategy of ecotourism ICDPs, the focus of the research, has emerged. The study also refers to the broader category of civil society on occasions when these normative allusions are significant.

Neopopulism

Examined in greater detail in Chapter 2, neopopulism refers to an emphasis on 'community' and on 'participation' in thinking on development and conservation respectively (Scheyvens 2002: 52–3; Potter *et al.* 1999: 68; Hettne 1995: 117). This is understood in opposition to a state of affairs that neopopulist critics regard as having characterised much economic development in the past and still today – a lack of regard for effects on local communities and a lack of community participation.

Positivism, functionalism and the advocacy of ecotourism

The ways in which to analyse a social phenomenon such as sustainable development, and the role of NGOs in advocating it, are the subject of ongoing debates influenced by varied perspectives. These debates draw on or run counter to previous discussions and paradigms in the social sciences. For this reason it is important to situate this study within the broader epistemological debate in order to make explicit the former's philosophical premises.

'What I want is Facts . . . Facts alone are wanted in life.' So said Mr Gradgrind in Dickens' *Hard Times* (Dickens 1981). For the positivist tradition, Gradgrind is right – facts come first, and conclusions can then be drawn from them. In broad terms, the positivist tradition dominated social research up until the 1960s, and remains very influential. Hughes describes succinctly the positivist version of social reality:

> The principal manner in which positivist social science constructs its version of social reality is by drawing a distinction between identifiable acts, structures, institutions, such as 'brute facts' or 'brute data' on the one hand, and beliefs, values, attitudes and reasons etc. on the other. These two orders of reality are correlated in order to provide the generalisations or regularities which are the aim of social life [. . .] In short, meanings are only allowed into scientific discourse if placed in quote and attributed to individuals as their opinion, belief, attitude.
>
> (Hughes 1990: 115)

The majority of research into sustainable development applied to tourism is broadly positivist. It generally attempts to provide factual information or guidelines that can then influence policy, or *post facto* assessments of how projects have fared measured against specific criteria. This tradition embodies a clear distinction between facts and values, viewing them as two separate spheres to be pursued sequentially (Finch 1986: 185). Such research is obviously of great service in the task of policy formulation. Yet there is more at issue in the debate over sustainable tourism than a disputation of facts (although these, too, are certainly disputed). Sustainable development remains a term that all can buy into, but for that very reason it may mask fundamentally different positions (Mowforth and Munt 1998: ch. 2; Pepper 1996: 74–5). Indeed, more facts and more studies have not necessarily brought sustainable tourism into sharper relief. If we regard as false the idea that there is a social reality that can be discovered independently of the vocabulary of society, then this state of affairs is less surprising.

My contention is that while the facts are subject to heated argument (as anyone who has read the differing assessments of high profile ecotourism projects or followed the academic debates will be aware), this obscures a general agreement on the values underpinning the advocacy of ecotourism. Either implicitly or explicitly, and in some cases in spite of authors' sceptical instincts, these values are assumed to be a normative goal for all.

It should also be remembered that even positivism embodies a commitment to the idea that society is ordered in a particular way – that social phenomena are essentially the sum of their observable parts (Hughes 1990). This can, and in the case of the study of ecotourism projects often does, tend towards a *presentism*, whereby *what is* comprises the focus for research while *what could be* (which clearly involves a leap of the imagination beyond the observable 'project') fails to be addressed. Many studies that declare projects a success (or a failure) based on a positivist methodology are limited in this respect.

This presentism is compounded by the functional emphasis of much written on ecotourism's role in rural development. Ecotourism is advocated on the basis that it reinforces cultural traditions that in turn promote conservation, that in turn reify those traditions (be it in a new way, backed up by small financial incentives in the form of ecotourism revenues), that then in turn promote conservation . . . and so on, in what its advocates would see as a virtuous circle. Following the Durkheimian tradition, its impacts may be regarded as strengthening important functions that maintain the community's way of life and a sustainable relationship with the environment.

For Durkheim, aspects of culture could be regarded as 'social facts' – facts rather than perspectives or opinions, hence assuming social agreement – a recognition of which enables policies that hold societies together (Morrison 1995). Aspects of culture are functional in this formulation – they serve a purpose in relation to a society *as it is constituted.* However, at the same time this functional approach rules out substantial social change to a society – again an emphasis on the present, *what is*, and a deprioritisation of *what could be* is the result. The conservative implications of this functionalist emphasis when discussing development in the poorest parts of the planet are clear.

It is precisely the functional emphasis in the advocacy of ecotourism (and the presentism that it supports) that deserves investigation, as it closes down the scope for thinking and debating on rural development. That ecotourism can constitute a 'win–win' solution, with mutually reinforcing benefits for a community and the environment, is presented as social fact by major conservation and development agencies. This 'fact' is backed up by research into monetary benefits and community input, as well as the internal logic of the argument that ecotourism can bring economic rewards premised upon conservation in a mutually reinforcing manner. Yet it may be less a *general* social fact, than the product of a *particular* social perspective. By asking questions beyond the project itself, we can reveal that to 'win' in one way (through ecotourism revenue) may also be to lose in another (to accept that development means maintaining a close relationship with the immediate natural environment). To 'win' in terms of jobs in conservation may mean having to accept that wider development, comparable with that in the donor countries, is off the agenda – especially when ecotourism is talked up as 'empowering', 'community' oriented and 'people first'. To accept that a localised symbiosis between nature and community, trumpeted as exemplary sustainable development, is a progressive innovation in rural development, is to deny the efficacy and possibility of the sort of thoroughgoing development that characterises the more wealthy economies.

A premise of this study is that orthodoxies akin to Durkheim's 'social facts' abound, but are only regarded as such due to the narrow parameters of the debates about rural development that characterise the discourse. In this sense this study's aim is to challenge the presentism implicit in much thinking on this issue.

Interpreting the debate

By contrast, this study attempts to *decentre* ecotourism as sustainable tourism. This refers to 'reject[ing] any form of empiricist epistemology in favour of an analysis of the structural relations and realities underlying the surface appearance of social and cultural phenomena' (Hughes 1990: 111). Put simply, it means placing the idea in a broader ideological and social context in order to better understand it.

Post-Second World War social science has diversified away from what historian E.H. Carr described in the classic *What is History?* as the 'cult of facts' (Carr 1990: 9), towards more interpretive approaches that can potentially scratch beneath the surface appearance of social phenomena. The modern interpretive tradition has always accepted that different view-points and sets of values come into social research. For one of its founders, Max Weber:

> [t]here is no absolutely 'objective' analysis [. . .] of 'social phenomena' independent of special and 'one sided' viewpoints according to which – expressly or tacitly, consciously or subconsciously – they are selected, analysed and organised for expository purposes [. . .] All knowledge of cultural reality, as may be seen, is always knowledge from particular points of view.
>
> (cited in Hughes 1990: 136)

One classic text in this tradition agrees, arguing that social science is 'a *selective* system of cognitative orientations to reality' (Parsons and Shils 1954: 167; my italics).

Within the modern sociological literature on tourism, this sentiment is echoed by MacCannell, one of the most influential figures in this field. Writing about modern tourism, MacCannell argues that authors should make explicit their ideological premises, and even commitments (MacCannell 1992: introduction). MacCannell's view is apposite. He argues that there are differences of values in a world of uncertainty and contested ideas, and that it is therefore appropriate to make explicit the premises of one's arguments in order to clarify the debate and help towards some kind of synthesis (ibid.).

This study is essentially interpretive – it attempts to interpret a phenomenon rather than to 'let the facts speak for themselves', Gradgrind style. As such, it is implicitly critical of a positivist approach to such investigations. A positivist approach to the issue may, for example, try to measure commitment to sustainable tourism against certain criteria set by NGOs, or a government body. However, the present study attempts to step back from such an approach and to analyse the meaning and social significance of sustainable tourism as formulated in the advocacy of ecotourism ICDPs. It may also shed some light on sustainable tourism's 'parental paradigm' (Sharpley 2000: 1), that of sustainable development.

The study draws on social constructionism. It attempts to analyse sustainability as a contested concept formed out of differing notions of how humanity relates or should relate to the natural world, rather than as a technical term. Sustainable tourism development resides on a view of the impacts created by tourism on natural environments and culture. Yet this is contested terrain. For example, what we regard as the 'natural' world, and the importance we ascribe to it, are historically and socially conditioned and not given. 'Nature' and 'culture' do not only exist 'out there' as objective phenomena, but acquire *social* meaning through the interaction of the human subject with the objective world (Urry and McNaghten 1998).

One author describes succinctly the process through which different versions of a concept can emerge and become self-reinforcing:

> [E]ach myth functions as a cultural filter, so that its adherents are predisposed to learn different things about the environment and to construct different knowledges about it. In this way beliefs about nature and society's relation to it are linked with particular rationalities, that support the modes of action appropriate for sustaining the myth.
>
> (Harrison and Burgess 1993, cited in Urry and McNaghten 1998: 4)

The research will try to make explicit the predisposition, or 'cultural filter', that leads the NGOs featured in the case study to equate ecotourism with sustainable tourism, in order to analyse this filter.

This is not to argue that positivist research into sustainable development is invalid – it is of course important to attempt to operationalise development, and this will involve measurement against performance indicators. However, this study contends that any such indicators are premised on particular conceptions of nature and development. It will be argued that there is little contestation of a distinctly ecocentric approach to development through ecotourism, in spite of the constant invocation of 'community', 'people', and 'empowerment'.

Establishing the discourse – a dual emphasis

The study is premised upon an analysis of the discourse advocating ecotourism as sustainable development in the rural developing world, and it adopts a dual emphasis to this end. First, the NGOs at the cutting edge of developing and implementing ecotourism in practice are clearly a key expression of this advocacy. Second, there is a wide range of academic and related writing that falls into the category of critical advocacy, and this provides a deeper focus on the concepts at stake. This dual focus provides the basis for a thorough examination of the discourse.

Whereas establishing a body of literature is fairly straightforward, the choice of NGOs is less so. It was necessary to establish a list (albeit not a definitive one) of NGOs which are engaged in the advocacy and/or the implementation of sustainable tourism development. There are many NGOs all over the world that are concerned, to a greater or lesser degree, and in a variety of ways, with tourism. Europe, North America and Australia seem to have generated a large number of relevant NGOs although, inevitably, given the international nature of development issues, environmental problems and tourism itself, many of the organisations concerned, even small ones, are international in scope.

The developing world is also considered to have a developing culture of NGOs (in the Philippines, for example, there are a considerable number of environmental and cultural NGOs, some concerned with tourism) (Princen and Finger 1994). However, as Princen and Finger have pointed out, southern NGOs have often been generated by, and depend upon, northern ones (ibid.). Also, it is the northern NGOs that are highly significant as aid donors, and the southern NGOs tend to be the recipients – it is in the developed world that the funding and the priorities for ecotourism ICDPs originates. It was therefore decided to limit the study to NGOs based in economically developed countries, as they are the agenda setters.

A list of NGOs was established on the basis of the author's knowledge, research through the internet and various publications. This list numbered some thirty quite diverse organisations, including: conservation NGOs such as the Aububon Society, Conservation International, WWF and Nature Conservancy; trade and certification organisations such as the International Ecotourism Society and Green Globe; think tanks including the Overseas Development Association and the International Institute for Environment and Development; campaigning bodies such as Tourism Concern and Survival International; and organisations with a human centred focus on development, well-being and human rights such as SNV and Oxfam. From this 'population' of relevant organisations, it was decided to choose a sample of case studies on which to focus the research.

Five cases were chosen purposively, based on three factors: the extent to which the organisation had pioneered ecotourism ICDPs and the amount of experience from this; in order to reflect the diversity of different types of NGOs within the broader population; and convenience, principally the availability of written sources on the subject. The first of these was considered to be the most important, as the study principally attempts to analyse an emerging strategy, that of ecotourism ICDPs, with the NGOs the instigators of that strategy.

The sample comprises:

1 WWF. WWF is the world's largest conservation NGO. It has been involved in projects that utilise ecotourism as a means of enabling its

primary aim of wildlife and nature conservation to be integrated with community benefits.

2 Conservation International (CI). CI is a large international conservation NGO. It was founded in 1990 on the basis of putting forward a critique of 'fortress conservation', and advocates 'community conservation'. It has pioneered ecotourism to achieve this end.
3 SNV. SNV is a Dutch based independent development agency with a focus on development in the rural developing world. SNV has pioneered ecotourism to engender sustainable development.
4 Tourism Concern. Tourism Concern is a vibrant UK-based campaigning organisation with an international profile, involved in promoting 'community tourism' in the developing world. Community tourism effectively constitutes ecotourism with a strong emphasis on 'community'.
5 The United Nations International Year of Ecotourism (IYE). This case study is quite different from the other four, comprising an international keynote event in which very many NGOs were represented, many of them prominent, including the four other individual NGOs featured as case studies. The author was aware of the event from early on in the research, followed it closely and has written about it elsewhere (Butcher 2006a, 2003b and 2003c). It was felt that this event was too important to exclude from the study, and therefore its extensive written output, and the debates around it, were treated as a case study alongside the individual NGO case studies. Had the research been principally about the NGOs themselves this would not have been feasible, but as the aim of the research was to reveal and analyse the assumptions underlying the advocacy of ecotourism as sustainable development, the documents arising from the event are clearly an important, perhaps the most important, example of this advocacy featuring the NGO sector.

Of course, a strong case could be made for other NGOs, such as the Audubon Society, the International Ecotourism Society and many others. Hence it is important that the case study organisations should broadly reflect the diversity of organisations in the population. An examination of the literature emanating from the population of NGOs enabled broad categories to be established, and then organisations were chosen reflecting the general trends in these categories. Two categorisations were drawn upon to inform this process.

The following list of categories, adapted from Doyle and McEachern (1998), proved useful in establishing common themes within the population of relevant NGOs, which could then be reflected in the sample of case studies chosen:

- their size;
- their principal aim;
- their character (e.g. campaigning, industry body, membership); and
- their relationship to the private and government sectors.

By briefly establishing broad answers to these questions for the population, a sample could be chosen that reflected both important common characteristics, and the diversity, within the population. Common characteristics of the case studies chosen, reflecting the population, include a diversity of funding sources, work with commercial, governmental and other NGOs, and international influence on the debates relating to development and conservation. Notably, the pervasive links between the NGOs and the other sectors – government and commercial – is a common theme.

Tourism Concern is distinctive within the case studies, in that it is a campaigning, membership organisation. However, it is the most influential of a network of similar organisations in Europe, and of other campaigns based elsewhere. CI and WWF are global conservation NGOs and, as such, represent the most important and largest section of the population. SNV is a rural development agency, with close links to the Dutch government, and hence reflects the development or 'well-being' organisations within the population.

The IYE involved many of the NGOs in the population, hence its importance as a case for this study.

While the above categorisation provides a series of fairly *technical* characteristics through which to assess the population and draw sample case studies, more important is to establish some parameters with regards to the organisations' *ideological* characteristics. This enabled cases to be chosen with divergent overall aims, reflecting those of the population, yet sharing an advocacy of ecotourism ICDPs. This is important, given that the population includes a variety of organisations whose primary stated aims range between development and conservation.

To achieve a clearer picture, the organisations were considered from the perspective of anthropocentrism and ecocentrism, the former referring to a more 'human centred' view of the relationship between humanity and nature, and the latter a greater emphasis on environmental preservation (Pepper 1996). With regards to the NGOs under examination, those whose primary aim is conservation would generally be regarded as having a significantly ecocentric outlook, and those with a greater emphasis on well-being and development are, to a greater degree, anthropocentric in their outlook.

Tourism Concern is a campaigning, membership organisation that prioritises community well-being and human rights. SNV is a development NGO, with an emphasis on rural developing world development issues. These two organisations could be regarded as more anthropocentric due to their well-being/development focus.

WWF is the largest global conservation NGO. Its primary aim is conservation although, as will be shown elsewhere, it, too, promotes its 'community' credentials. CI is a global conservation NGO also. However, it is distinctive in that its emergence was the product of a split within a major American conservation NGO over the issue of the need to involve communities in conservation and deliver benefits for those communities.

Thus, it provides an insightful example of the development of ecotourism as a tool for community-based development and conservation. These two organisations could be regarded as lying more towards ecocentrism, having their roots in conservation.

Finally, the UN IYE drew on a range of NGOs, as well as governmental agencies. Therefore its keynote documents comprise a synthesis of ecocentric and anthropocentric perspectives.

The diversity of NGOs involved in some way with ecotourism is striking – volunteer groups such as Voluntary Service Overseas (VSO), think tanks such as the IIED, development NGOs such as SNV, campaigns such as Tourism Concern and conservation NGOs such as WWF to list just a few. However, the five cases provide sufficient breadth while enabling a depth of analysis into the discourse under investigation.

Analysing the discourse

Written documents and accounts have been taken as the principal evidence – it is here that ecotourism is proposed, rationalised and discussed. Such documents can be seen in two senses. They are written statements of facts, opinions from or about an organisation or, in the case of the IYE, organisations. They reflect the organisations' outlook and perspectives. However, '[d]ocuments [. . .] do not simply reflect, but also construct social reality and versions of events' (May 1993: 138). In this case, they are part of producing and reproducing a discourse about sustainability. May argues that documents are therefore mediums through which social power is expressed (ibid.). Thus researchers need to bring to bear a critical eye, one which attempts to put the documents within their cultural context (May 1993: 138–9).

The study does not include a formal textual analysis of the material produced by the various NGOs. This would end up assigning a mechanistic way of thinking to often complex and abstract issues. However, an analysis of ideas expressed through these documents is essentially a discourse analysis. In this context, a discourse refers to 'a set of meanings, metaphors, representations, images, stories, statements and so on that in some way together produce a particular version of events' (Burr 1995: 48).

Discourse analysis is associated with a number of qualities that this study seeks to emulate. First, discourse analysis takes a 'critical stance to "taken for granted knowledge"' (Burr 1995: 35). This is apposite here. As will be argued more fully elsewhere, the consistent association of ecotourism with sustainable development (or as having the potential to move towards this, especially when compared with other development options) has tended to grant its assumptions the status of 'taken for granted knowledge'. It is these assumptions – the efficacy of local community participation, the desirability of development based upon traditional culture and the importance of environmental fragility – that are examined in succeeding chapters.

Ecotourism ICDPs as sustainable development could be regarded as a paradigm in rural development, or as part of the sustainable development paradigm – a paradigm being a set of widely supported and mutually reinforcing assumptions underpinning something taken to be true or normatively good (Howarth 2000: introduction). Its self-image is as an 'alternative' paradigm – as part of the Alternative Development Paradigm (Pieterse 1998). It is certainly a viewpoint that has emerged and become established in opposition to previous modernisation-inspired notions of development. However, within influential sections of the NGO and academic community it has become an orthodoxy itself (ibid.).

Sustainability, a goal that ecotourism is consistently associated with, has certainly achieved paradigmatic status (Pepper 1996: 260). In fact, it is remarkable how rapidly the term 'sustainable' has become a prefix for other terms – development, tourism, housing and communities to name a few. Its paradigmatic status may be warranted given the extent of ecological problems that have come to the fore in recent times. However, it is also possible, as two pioneers of political ecology put it, that such a dominant paradigm can become like 'spectacles' (Briggs and Peat 1985: 24–34). To develop their analogy, the spectacles are donned by researchers examining the subject, and enable them to focus clearly on a certain range. However, the spectacles may be insufficient to see further. Moreover, they may blur objects right under the researcher's nose. The aim of this study is not to ignore mainstream conceptualisations of sustainable development, but to remove the 'spectacles', in order to better understand its premises and implications in a specific case.

The association of sustainable development with linking conservation and development could even be described as an *ideology*. Milton describes the role of an ideology in relation to action succinctly. It 'fulfils both cognitive and practical functions; it enables people to understand the world and their place within it and forms a basis for action' (1996: 83). Hence, it 'legitimates and justifies' courses of action (ibid.: 82). In so far as the assumptions underlying an ideology are unchallenged, it can constitute a self-referential and self-reinforcing way of looking at the world, one that, in this context, is passed around funding agencies, academics, students and southern NGOs. In this sense, this study is a critique of the terms of reference of the debate.

Discourse analysis is one way of interrogating and challenging dominant paradigms and widely held ideologies. It shows us that there can be systems of knowledge constituted around certain accepted propositions, that produce 'truths', or accepted parameters for debate. This proposition is at the heart of the analysis. Also, while discourse focuses on the expression of ideas (in speech or written documents) as opposed to actions, discourse itself is an intervention in the social world, a form of social action (Burr 1995). Knowledge and social action go together – different constructed versions of reality invite different practical solutions (ibid.). It is in this straightforward sense that discourse analysis is adopted in this study.

However, discourse analysis is also often associated with post-structuralist thought, and most notably with Foucault, who exerts considerable influence in this field. Although Foucault's notion of the archaeology of knowledge is implicit in this study, in that it tries to locate ecotourism ICDPs in the trajectory of thinking on development and conservation, his general understanding of discourse is not one shared here.

Foucault's disciples tend to see social reality as essentially perspectival, and ideas as constructions of particular sets of social experiences and premises. I would counterpose a humanist view – it is possible to attempt to speak from a common, human perspective, and engage in a battle of ideas on that basis. In development, the common humanist aspiration to make available modern technology and the benefits of modern science to all has tended to be reined in by a post-structuralist influenced emphasis on cultural differences and different 'knowledge systems' (see Chapter 5). Notably, in the advocacy of ecotourism ICDPs in the developing world, the role of traditional culture is accentuated, not as a stepping stone towards greater development, but as part of a localised, steady state between the community and the environment. It is precisely this relativisation of development that the study seeks to challenge.

The wider advocacy of ecotourism

The documents from the NGOs are obviously a key element of the study – it is the NGOs that are at the cutting edge of developing and implementing the sort of project under analysis here. However, throughout, the study adopts a dual focus and also considers the wider advocacy of ecotourism in exemplary academic and related literature. Ecotourism has very many supporters here. Sometimes the extent of agreement around the ideas is masked by quite intense debates about ecotourism's merits, but as argued in Chapter 2, these are debates *within* the advocacy of ecotourism – they comprise critical support, but support nonetheless. Analysis of the wider literature enables a more thorough exploration of the assumptions and positions which, while clear in the NGO related documents, are developed in this literature.

Hence, the study attempts a critical synthesis between an important strand of literature on development and conservation on one hand, and the specific literature on ecotourism from the NGOs and elsewhere on the other.

Finally, as the reader will have gathered by now, this study is not a detached analysis, but is explicitly an intervention into the discourse. This is true of much writing in the field of development (Milton 1996: 74), and of the debates about the merits of ecotourism (Butcher 2003a). There is a great deal of advocacy of ecotourism in a range of texts and academic papers and commentaries on ecotourism's development potential. Often this advocacy is low key – the uncritical acceptance of ecotourism as exemplary sustainable development in rural areas means that it need not be anything else. Yet even apparently straightforward positivist accounts of the relative success or failure of projects (of which there are many) adopt criteria that themselves reflect a particular standpoint on development and conservation – they implicitly adopt a stance on the social world beyond their study. The aim of this book is to make explicit the premises of ecotourism's advocates, and present an alternative to a development agenda that ties rural communities to localised natural limits, and talks this up as 'sustainable development'.

2 Ecotourism in development perspective

Introduction

This chapter locates ecotourism in post-war thinking on development and conservation, and hence provides a basis for understanding its advocacy as sustainable development. In keeping with the aims and objectives of the study, the chapter focuses on the broad conceptual context of ecotourism ICDPs, rather than operational issues. If the latter were the focus, then a review of reports from the field would be appropriate. However, in order to establish this *conceptual* context, the chapter identifies and synthesises two interrelated strands of literature – on conservation and on development – that together provide the basis for the claims made for ecotourism. The chapter also attempts to situate the study in the context of exemplary literature on ecotourism.

The chapter is divided into three sections. In the first section it is argued that ecotourism reflects a coming together of important strands of thinking on development and on conservation, concepts traditionally seen as being at odds with each other. As grand, modernist views on development have been subject to criticisms, development thinking has adapted to a 'community' agenda. Conservation thinking has also tended to gravitate towards this agenda, in part as a response to criticisms of 'fortress conservation'.[1] The resulting convergence is reviewed here, as it provides the context for the development of ecotourism ICDPs, which are, in effect, manifestations of this convergence – the argument commonly made is that ecotourism constitutes sustainable tourism development as it can bring development *on the basis of conservation* (Goodwin 2000: 97–112). This part of the review also briefly considers the post-modern character of the critique of the modernisation paradigm, and also the 'greening of aid', which refers to funding for development initiatives that reflect the aforementioned convergence.

In the second section, the chapter examines the outcome of this convergence – a dual emphasis on *neopopulism* and *civil society* in an important strand of discourse on development. Neopopulism well describes the result of the convergence between conservation and development – it refers to an emphasis in development on community-based action and participation, which in turn is central to the advocacy of ecotourism in the case studies. Also, the NGOs featured in the case studies all lie within the broad category of 'civil society'. Civil society organisations have gained in prominence as organisations able to operationalise the neopopulist agenda referred to above. Therefore the review will look at the concept of civil society in order to inform a clearer understanding of the trajectory of ecotourism as a development strategy.

Put simply, *convergence in development and conservation thinking* has resulted in the growth of a *neopopulist* agenda, an agenda carried through by *civil society* organisations (NGOs). These three aspects provide a conceptual context for the development of ecotourism ICDPs, and thus comprise the basis of parts one and two of the chapter.

The third section considers the study in the context of other critical literature on tourism and development. It also serves to demarcate this study from other critical studies and commentaries.

In conclusion, the chapter draws the various ideas together, and establishes how they set the parameters for the analysis in this study.

The coming together of conservation and development

This section establishes the context within which the discussions about tourism's role in sustainable development in the developing world takes place. One way of simply but usefully conceptualising the shift in post-Second World War development thinking is to recognise that there has been a growing critique of 'development from above' and a promotion of 'development from below'. Parallel to this there has been a critique of 'fortress conservation', which could be characterised as 'conservation from above' and a promotion of local participation in conservation, which in turn could be regarded as 'conservation from below'. In both fields there has been a tendency to give increased importance to participation by the local communities concerned. The imperative to change from 'development from above' is captured in the following quotation:

> Surely, if decades of failed international development efforts have taught anything, it is the folly of induced, uniform, top-down projects. Such schemes ignore and often destroy the local knowledge and social organisation on which sound stewardship of ecosystems as well as equitable economic development depend.
>
> (Rich 1994: 273)

A similar sentiment is also evident with regard to 'fortress conservation'. Adams argues that, on the part of conservation thinking: '[t]here has been a self-conscious effort to move beyond environmental protection and transform conservation thinking by appropriating ideas and concepts from the field of development' (Adams 2001: 3). In this fashion, conservation and development thinking have tended to converge, and the result of this has been the emergence of common themes. Prominent among these are the emphasis on community development and participation, and the overarching imperative to aim for development that is deemed environmentally sustainable.

Conservation: from fortress conservation to community participation

The key driver for the shift from the perspective of conservation in the developing world has been a tension between the conservation agenda as envisaged by Western conservation organisations and the development aspirations of developing world countries (Doyle and McEachern 1998; Mowforth and Munt 1998; Preston 1996: 306; McCormick 1995: 99; Redclift 1990). These tensions forced or encouraged conservationists to address development as a way of making conservation feasible in the context of poverty and a dearth of development (Ghimire and Pimbert 1997).

We can trace this tension in the main post-Second World War codifications of conservation thinking. In this period, global conservation initiatives have provided an arena for disagreements over the relative priority for environmental conservation in the developed and developing worlds, and have witnessed changes in conservation thinking broadly towards a greater recognition of development needs (ibid.).

An important early example of this is the Man and the Biosphere (MAB) programme, which was launched from the Intergovernmental Conference of Experts on a Scientific Basis for Rational Use and Conservation of the Biosphere, a UN initiative held in Paris in 1968 (Adams 2001: 49–50). This programme exemplifies the growing links being made between conservation and development needs (Blaikie and Jeanrenaud 1997). With reference to MAB, Adams points out that conservationists 'realised increasingly through the 1960s that they could not influence decisions about the use of natural resources in the Third World *unless they were at least prepared to talk in the new language of development*' (Adams 2001: 49; my italics).

MAB's aim was to 'develop the basis within the natural and social sciences for the rational use *and* conservation of the resources of the biosphere' (Gilbert and Christy 1981: 710; my italics). Specifically, the remit was to look at the *relationship* between 'natural' ecosystems and 'socio-economic processes' (Adams 2001: 50) in order to secure benefits for both conservation and well-being. Adams also makes the point, however, that MAB was primarily concerned with conservation, although it dressed up conservation in what he describes as the 'new clothes' of human ecology[2] (ibid.). An example that emerged from MAB, apposite for this study, is the development of biosphere reserves – zoned nature reserves aiming to conserve biological diversity and the genetic information contained within. These were conceived of as having the facility for human activities to continue on the periphery of the reserves. ICDPs have emerged as just such a human activity, permissible on the periphery, to aid conservation and human communities. In alluding to human needs in this way, the establishment of biospheres clearly placed those needs behind conservation in the list of priorities in these particular areas by logically deprioritising the development question, and limiting it spatially to the periphery.

Referring to biosphere reserves developed with such priorities, Batisse writes that:

> Experience already shows that when the populations are fully informed of the objectives of the biosphere reserve, and understand that it is in their own and their children's interests to care for its functioning, the problem of protection becomes largely solved. In this manner, the biosphere reserve becomes fully integrated – not only into the surrounding land use system, but also into its social, economic and cultural reality.
>
> (Batisse 1982: 107)

According to this formulation, the integration of human concerns into conservation should take place on the basis of consultation and the delivery of some benefits to the populations concerned. Based on a recognition of the benefits, these populations may concur with the conservation of biosphere reserves. It is central to the advocacy of ecotourism that it can play a role in delivering benefits and incentivising conservation in this way (e.g. Fennell 2003; Goodwin 2000; Ziffer 1989). However, it is also clear that conservation is the driving imperative here, and attempts to benefit the population are largely moulded around this priority. Colchester has written about the negative consequences of this approach for the Masai in Kenya, who were promised benefits, including those arising from ecotourism, in return for restricting their activities to the 'buffer zone' of the Amboseli National Park (1997). The benefits proved illusory, and the Masai were effectively excluded from compensating for this by seeking a livelihood within the park (ibid.).

This view of conservation being integrated with development was fostered by MAB in the 1970s, and has developed through the UN Conference on the Human Environment held in Stockholm in 1972, the publication of the *World Conservation Strategy* (WCS) (IUCN *et al.* 1980) and *Our Common Future* (WCED 1987), the staging of the 1992 UN Conference on Environment and Development in Rio (UN 1993) and the Johannesburg World Summit on Sustainable Development ('Rio plus ten'). It remains influential in key codifications of conservation. Indeed, the example of biosphere reserves in particular is apposite to this study, as biosphere conservation is the basis for WWF and CI, and a host of other conservation NGOs, to be involved in ecotourism – many ecotourism projects are in biosphere reserves. The other case studies featured share this concern with biodiversity conservation, though it is less central to their *raison d'être*.

In 1972, the UN Conference on the Human Environment, often considered the first global conference to highlight environmental threats, featured scepticism on the part of developing world representatives as to the motives of conservationists who, in the former's eyes, sought to cap development in the developing world in the name of conservation (McCormick 1995: 99). This did not fit well with the ambitious development plans of many newly independent post-colonial developing world governments, which naturally aspired to move away from 'third world' status (ibid.; Adams 2001; Preston 1996). Post-independence, national parks and reserves were generally not a priority, for example, for newly liberated African states that sought to adopt wide-scale economic development. Such reserves were often associated with the previous colonial regimes that had established them for pleasure and out of a conviction that Africa was, and should remain, a place of wilderness (Leech 2002; Reader 1998). However, at this time some Western environmentalists tried to encourage the developing world to see 'the virtue of living off the interest of their natural resources, not the capital' (McCormick 1995: 49), and there was some unease at the prospect of post-colonial regimes following what many environmentalists perceived to be environmentally damaging large-scale industrialisation (ibid.). Such tensions characterised the emerging discussion of 'sustainable development', discussions often seen as originating at Stockholm (ibid.; Hall 1998).

The WCS (IUCN *et al.* 1980) addressed the question of development, but grew from a determination within the WWF and the International Union for the Conservation of Nature (IUCN) to find new types of development that were more environmentally benign (Adams 2001: 270). The WCS in many ways marked a maturation of the environment/development debate. It argued that, in fact, development could be reconfigured to *promote* conservation and that, rather than local people paying a price for conservation, they could benefit from it (ibid.). This is clearly the basis for the rationale behind the ICDPs, which seek to implement the philosophy of the WCS at a local level within rural communities. The change in thinking marked out by the WCS was considerable (ibid.). As applied to rural areas, principally in the developing world, we could characterise the change as a move from 'fortress conservation' (Adams and Hulme 1998) to 'community conservation' (Scheyvens 2002: 88–93; Adams 2001: ch. 12; Pimbert and Pretty 1997). In similar vein, ecotourism ICDPs are sometimes referred to as 'community based natural resources management' (e.g. SNV/Caalders and Cottrell 2001: 19 and 22), or simply as 'community based ecotourism' (e.g. Scheyvens 2002: 71), reflecting this changed thinking.

Our Common Future, published in 1987, is notable in that it is the first time the UN General Assembly had explicitly discussed environment and development as one single problem (WCED 1987: 3). Ostensibly the report does not approach the issues from a development or a conservation perspective, but from the perspective that the two are *inseparable*. This perspective informed the development of ICDPs, including those involving ecotourism, the central argument for which is that development and conservation are, or can be, 'symbiotic', or at least brought into some sort of mutually supporting balance (McShane and Wells 2004; Ghimire and Pimbert 1997).

The Stockholm *UN Conference on the Human Environment*, the *World Conservation Strategy* and *Our Common Future* reflect the drawing together of development and conservation. However, in a sense they also reflect the tension between these two categories rather than a resolution of that tension. For example, Adams (2001: 73) argues that *Our Common Future* was development dressed up to look environmentally sensitive, whereas the *World Conservation Strategy*, produced by the IUCN and WWF in 1980, was essentially the reverse, environmentalism dressed up to be acceptable with regard to development. Codifications of sustainable development both prior and subsequent to these examples have always produced critics on either side: conservationists who believe development has too much priority; and those favouring development who see conservation as constraining. Sustainable development has not resolved this tension (Adams 2001; Redclift 1990).

The most important manifestation of the coming together of discussion on development and the environment is the UN Conference on Environment and Development of 1992, often referred to as the Rio Earth Summit (UN 1993). The summit effectively launched sustainable development as a new rhetorical orthodoxy, and attempted to move towards a new development, one that was more benign towards the environment and hence more 'sustainable' (Baker 2005; Adams 2001; Mowforth and Munt 1998).

However, the preparatory commissions for the conference revealed that strong divisions between the developed and developing world countries remained (Adams 2001: 80; Mowforth and Munt 1998: 22; Pearce 1991: 20). Throughout the Rio Earth Summit process there were important differences over the key problems that were to be addressed. Developing world governments expressed worries that agreements on the environment would prove to be restrictive to growth and to their freedom to use their natural resources to best economic effect (Adams 2001: 83). The opposition to Western priorities was in part motivated by a sense that the process was hypocritical in its calls to preserve biodiversity in the developing world, when the developed world had become developed precisely by clearing forests and transforming their environments in the course of the development of agriculture and industrialisation. In fact this latter point was made strongly in 1990 in the preparatory discussions for the UN Conference on Environment and Development by a group of developing world states calling themselves the G77, who made it clear to the G7 (the caucus of the seven leading industrialised nations of the time) that they would oppose a global forest convention on the basis that it would be restrictive for them, but much less so for developed nations who had long since levelled the vast majority of their forests (Adams 2001: 89).

To placate the fears of developing world representatives, they were promised new financial resources and technical assistance; this was seen as a development pay off for signing up to global sustainability (Jordan and Voisey 1998). However, since the Earth Summit, aid budgets have tended to fall, and very few Western countries outside of Scandinavia have got close to the 0.7 per cent of GDP baseline figure that was signed up to (ibid.). One could argue that this crisis in development aid should be viewed as logically, if not morally, *prior* to a discussion about altering international priorities in favour of the environment. The reality is the opposite – the 'greening of aid' has taken place alongside the decline of aid budgets relative to donors' GDP.

Factors such as the Aids epidemic, global economic trends and the relative dislocation of parts of the developing world from the world economy have set back the economic prospects for many of the poorest countries in the world. In this context the talking up of green, 'capital cheap' projects as sustainable development for impoverished rural communities can seem perverse.

Development: from modernisation theory to sustainability and the greening of aid

ICDPs do not just represent a move within conservation thinking to integrate development goals. There is also a move in the opposite direction, from the debates around economic development, towards the need to factor in conservation. This latter trend is highly evident in the trajectory of post-Second World War development studies, and most clearly evident in the move from traditional development, conceived of as principally economic in character, to the 'triple bottom line' of economy, environment and culture that is central to contemporary codifications of sustainable development (Muller 1994).

From the 1950s to the 1970s, modernisation theory characterised much development thinking (Preston 1996: Ch. 9). Modernisation theory held that economic growth led to, indeed constituted, development (ibid.). Also, it was assumed that there is a unilinear path to development that all countries, prospectively, could follow. Such theories draw on the Enlightenment influenced conception of human progress, that the progressive ability of humanity to harness nature and, through this, develop economically, constitutes development (Preston 1996; Hettne 1995).

Archetypical of post-Second World War modernisation theory were the ideas of US economist Walt Rostow, who saw the undeveloped countries as needing to pass through five stages in order to attain the sort of developed status enjoyed by the United States and the rest of the developed world (Potter *et al.* 1999; Preston 1996: 175–7; Ghatak 1995). The stages were common for all countries, and the highest stage was the consumer capitalism of post-Second World War America.

However, equally, the emphasis on extensive economic growth in the former Soviet Union can also be described as exemplary of the modernisation paradigm (Potter *et al.* 1999; Preston 1996). Although Rostow was an advocate of the American path to development, and this was in opposition to the Soviet Union's development efforts during the cold war, each side in the cold war shared a common faith in the power of economic development, or the raising of the forces of production, to deliver to mankind. Indeed, some regard Rostow's ideas as part of cold war ideology, as they presented to a dangerously unstable developing world of the post-Second World War decades the possibility to develop dramatically through remaining within the orbit of the capitalist world (Preston 1996: 177). Both systems, capitalist and communist, saw large-scale production and industrialisation as key to development at home, and also, prospectively, abroad (Potter *et al.* 1999).

However, from the 1970s, the modernisation paradigm was challenged by the emergent critique of development thus constituted (Narman and Simon 1999; Potter *et al.* 1999; Hettne 1995). The emphasis on economic categories such as GDP per capita, and the view that different countries would follow a common path, were increasingly questioned (McMicheal 2001; Hettne 1995). Over time, *sustainable* development emerged as a new orthodoxy, an orthodoxy that all buy into, but on which there remains much disagreement (Adams 2001; Redclift 1990). Sustainable tourism development is the offspring of its 'parental paradigm', sustainable development (Sharpley 2000: 1), and it, too, remains hotly contested. This study looks at the conception of sustainable development applied by a number of NGOs to the developing world. It is in effect a study of one manifestation of this new development paradigm.

One important part of the challenge to the modernisation paradigm has been the increase in concern over the environmental effects of growth, manifested in the elevation of the environment to a key issue in contemporary politics (Adams 2001; Urry and MacNaghten 1998). The clearest manifestations of the growing concern with the environment has been the growing prominence of environmental issues in public consciousness, and this in turn has fuelled the growth in membership of, and support for, NGOs which address such concerns (Hilhorst 2003). One could include in this the success of NGOs concerned with the environmental and cultural impacts of the growth of tourism (Mowforth and Munt 1998: 157–62), including those featured in this study. Tourism Concern, and similar campaigns in many countries around the world, are pertinent in this respect – they reflect a significant public disillusionment with modernity in the form of mass commercial tourism.

Post-modernity and development thinking

The rejection of the modernisation paradigm by many has involved the growth of what can be regarded as post-modern thinking on development. Indeed, post-modernism has had a profound influence on development discourse. Post-modernity rejects the idea that there are common standards for development, and takes diversity, and hence diverse forms of development, as its starting point. It also shares much common ground with neopopulism (Scheyvens 2002: 36–7). Both reject modernisation, embrace diversity, support local action and respect local voices, recognise that power relations inform the construction of knowledge and the establishment of research agendas and development priorities, reject the notion of a single truth, and accept that the meaning of development is contested and subjective (Blaikie 2000).

Also post-modern thinking tends to be *against* aspects of modernity (Therborn 1995). As grand schemes for general development have appeared less tenable, particular, small-scale projects have acquired greater purchase. These projects often elevate local knowledge and skills, and eschew the transformative agenda of industry on a large scale. The criticisms of the mass tourism industry, and the talking up of ecotourism, are exemplary of this trend. In this vein, the advocacy of ecotourism as a development tool often explicitly criticises mass tourism as an exemplar of modern development, and counterposes ecotourism as a more benign option (Butcher 2003a).

Friberg and Hettne (1985: 231), advocates of the post-modern development agenda, argue that capitalism and state socialism were based on rationality, growth, efficiency, centralisation and large-scale production. These features, they argue, have pushed to the margins the interests of local communities and environmental sustainability. In their place these authors argue for small-scale, decentralised, locally oriented development and a sceptical attitude towards rational planning.

One could question the basis of their argument. Many who have written critically about capitalism have based their critique precisely on its *irrational* character. For example, in relation to the operation of international money markets, and through the inevitable duplication of effort in research between competing companies, capitalism has been criticised as essentially irrational (Monbiot 2004; Callinicos 2003). Also, many who have written about the planning mechanisms developed by the former 'communist' countries in eastern and central Europe have pointed out the irrational, inefficient nature of these. State bureaucrats had an incentive to meet plan targets, but little to meet them with any level of quality, and hence planners would often try to put forward low targets, targets that could be achieved more easily (Lane 1984).

But if Friberg and Hettne are right, then it is legitimate to ask what is wrong with rationality, growth, efficiency, centralisation and large-scale production? Logically, if we move from rationality, we move towards irrationality, and if we move from efficiency, we move towards inefficiency. Large-scale production can confer economies of scale, irrespective of social system, and potentially the production of more with less. It is hard to accept these authors' wholesale rejection of modern development, and their advocacy of small-scale, local development in its place. However, it will be argued that this anti-modern viewpoint is clearly evident in the advocacy of ecotourism as sustainable development (Chapters 5, 7 and elsewhere).

The alternative of Friberg and Hettne (1985) is 'endogenous development', which is rooted in local communitarianism and self-reliance, and is limited by both globally and locally defined environmental limits. In seeing communities as rooted in their specific relationship to the land, these authors effectively *naturalise* social relationships, and even adopt the term '*natural* communities' (my italics) as the unit at which development should be organised (ibid.). This theme, as it pertains to the advocacy of ecotourism, is revisited in Chapter 7.

Friberg and Hettne put forward a conception of the social world that is rooted in local ecology. People have agency within society – they can make and remake themselves and their society – but at a local level, and within boundaries that are determined not just by natural processes globally, but by the local environment into which they were born and in which they live. This rooting of the community's agency in the specific relationship with their immediate natural environment is a feature of the advocacy of ecotourism, as will be alluded to throughout the analysis (Chapters 4–7).

If we compare this conception of the social world with the development of human societies historically, it appears incredibly limited and limiting. All human history has involved the struggle to progressively harness nature for human ends, and has involved migration and experimentation in search of better ways to live and a more rational understanding of nature. Yet the 'ecodevelopment' philosophy put forward by Friberg and Hettne (1985), which is, it will be argued, characteristic of ecotourism (Chapter 7), restricts discussion of social change and, by implication, reduces the scope of human agency to what is local and 'natural'. It is a clear example of the anti-modern character of post-modern thinking in this field.

Indeed, these authors are clear that their preferred mode of development draws on pre-capitalist societies, and rejects the modern state in favour of networks and voluntary organisations and 'non party' politics (Friberg and Hettne 1985: 237). Again, this, as this study shows elsewhere, is characteristic of the advocacy of ecotourism as sustainable development. This study takes a critical stance with regard to such post-modern thinking on development. It will be argued that the denigration of the modern and the elevation of the pre-modern in the form of 'traditional knowledge' has little to offer rural developing world societies (Chapter 5).

The greening of aid in development thinking

The shift in development thinking towards sustainable development is sometimes regarded as 'new labels on old bottles' (Hall 1998: 13). Some critics generally (Adams 2001: ch. 4), and with specific regard to sustainable tourism development (Wheeller 1992), dismiss the rhetoric of sustainable development as lacking in substance. However, there is evidence to suggest that very real changes have resulted from the new school of development thinking.

Notably, the 1980s and 1990s have witnessed a new policy agenda that promotes the factoring in of environmental considerations to overseas aid

(Ghimire and Pimbert 1997; Robinson 1993). This new agenda has been described as 'Green Development' (Adams 2001), and elsewhere as premised on the 'greening of aid' (Solesbury 2003). The latter term gets across well the coming together of aid, traditionally associated with economic development, and the growth of environmentalism's influence in political and social debates.

The greening of aid refers to increasing amounts of aid being channelled through NGOs into the developing world, often at a local level, rather than from one (developed world) government to another (developing world) government. Such aid aims to meet the priority of 'sustainable livelihoods' (Solesbury 2003). This phenomenon has broadened the role of NGOs such as WWF, CI and SNV, which have received government funding connected with these priorities. The greening of aid has also created channels of funding for community tourism projects championed by Tourism Concern, most prominently in their *Community Tourism Guide* (Tourism Concern/Mann 2000). The UN International Year of Ecotourism, and the related documents (UNEP/WTO 2002a and 2002b), are also very much a result of the growth of ICDPs, which in turn is premised on the greening of aid.

The greening of aid is usually identified with the 1980s. There have been various manifestations of it, most notably changes in aid policy from major supranational and national aid donors who have been taking on board the increasingly prominent environmental agenda (Ghimire and Pimbert 1997). For example, the World Bank put on the record its first official commitment to linking its work with environmental considerations in 1984 (Adams 2001: 327). The guidelines laid down included that the bank would not finance projects that 'cause severe or irreversible environmental degradation', or that would 'significantly modify' biosphere reserves, national parks or other protected areas (ibid.). Such places have subsequently become the focus for ecotourism ICDPs.

Environmental groups have maintained pressure on the World Bank and other institutions concerned with global development, and the World Bank has responded to this by making further changes. In 1987 the World Bank created its own environmental department, with forty new staff, and new scientific and technical staff in its regional offices (Adams 2001). Attempts were made to formally factor the depletion of non-renewable resources into project appraisals. The establishment of the Global Environmental Facility (GEF) in 1991, working through the United Nations Development Programme (UNDP) and the UN Environment Programme (UNEP) as well as the World Bank, is also significant – it is involved in funding ICDPs, including those involving ecotourism.

National, as well as international, agencies concerned with aid have also adapted to the greening of aid. For example, in 1997 the UK Department for International Development (DfID) produced a White Paper committing them to the promotion of 'sustainable livelihoods' and to the protection and improvement of the 'natural and physical environment' (DfID 1999b).

The relationship of NGOs to the process of the greening of aid is a two-way process. The large international conservation NGOs, such as WWF, played a central role in coordinating pressure on the World Bank and development agencies such as DfID. They could also claim to be representative of a strong and growing strand of environmental concern evident in, for example, the increasing number, support for and membership of environmental NGOs themselves (McCormick 1995). However, the taking on board of the greening of aid also created a space for the NGOs to become more involved in distributing aid, acting in a consultancy role, or simply pressurising donor agencies to fulfil their stated environmental claims. In this way the NGOs have been important in shaping the greening of aid, and hence also the environment within which they have been able to develop into actors concerned with development as well as conservation.

Some remain unconvinced as to the extent to which the greening of aid has transformed development practice (Fox and Brown 1998). However, it is likely that as well as adapting to external pressure, the development agencies have also internalised the new philosophy of sustainable development and also community development involving integrated conservation and development. Indeed, the growth in interest in ecotourism from donors is indicative of this (Leijzer 2002; Sweeting 2002; Woolford 2002). Ecotourism ICDPs, the principal approach of the NGO case studies, and a plethora of other NGOs, in their efforts to move towards sustainable tourism development in the developing world, are limited evidence at least of the greening of aid in practice.

Characterising the convergence – neopopulism and civil society in the advocacy of ecotourism

Neopopulism

The chapter has thus far established that important strands of development and conservation thinking have converged around the themes of small-scale community-based development, environmental sustainability and empowerment of the community. These themes are profoundly *neopopulist* in character. The next section establishes and critically reviews the main parameters of neopopulist thinking on development. The conceptualisation, and criticisms, of neopopulism referred to here provide a context for the more specific analysis of ecotourism in subsequent chapters. As such, the study is a critique of the claims made in the advocacy of ecotourism as sustainable development, and also a limited critique of neopopulism.

From populism to neopopulism

The derivation of the term populism is enlightening in understanding contemporary neopopulism. Populism entered the political vocabulary with the formation in 1892 of the Populist Party in America (Marshall 1998: 508). Today, the term is most often used to describe any political movement seeking to mobilise people as *individuals*, rather than as members of a *particular socio-economic group*, against an overbearing authority, usually the state or foreign states (ibid.). Also, populism has been associated with the expression of grievances against the free market and perceptions that it works unfairly against a spatially defined section of 'the people'. This was the case with American farmers in the late nineteenth century and is a relevant perspective today. More recently, populism has been associated with Peronism in Argentina and with the popular opposition to communist rule in the former Eastern bloc (ibid.). As such, populism has an affinity with 'the people', and also with the notion of civil society – a realm of social action separate and distinct from the state and market.

Neopopulism is a broad term, open to different interpretations, but it relates closely to populism as outlined above. There is consistently an emphasis on the *local*, the *community* and their *control* over their own distinct development in the face of the market, state and supranational bodies such as the World Bank and International Monetary Fund (Potter *et al.* 1999; Hettne 1995).

The NGOs in this study draw heavily on neopopulism. There is a great emphasis on participation and on 'bottom up' planning in their literature. This is mirrored in the general literature on tourism, for a large section of which 'community consultation', 'empowerment' and 'participation' has become something of a new rhetorical orthodoxy (Chapter 4).

It is not the intention here to provide an in-depth exposition of neopopulism, but rather to establish its key ideas. It is argued that these ideas strongly inform the approach in the case studies to sustainable tourism in the developing world. Thus, this section explores the character of neopopulism through key literature, establishing that it represents a critique of, and a retreat from, the 'top down' development strategies often associated with the modernisation paradigm. It will also introduce 'community' as the key legacy from neopopulism to the NGOs, in the latter's approach to sustainable tourism in the developing world. The related issues of control, participation, community and scale are examined in greater depth, in relation to the neopopulist approach of the NGOs.

Neopopulism in development can be traced back to the 1960s. In this decade the Dag Hammarskjöld Foundation (DHF) was discussing what it termed 'another development', one oriented towards local communities (Potter *et al.* 1999: 67). Central to the DHF was a promotion of self-reliance within a community, endogenous (as opposed to exogenous) development and the establishment of basic needs and participation, not as a means to modern development at some later stage, but as an end in themselves (ibid.). That development should be endogenous was a key argument of the foundation. This is often referred to as development in the sense of being 'what people do for themselves' as opposed to 'what is done to them'. The latter, of course, was associated with the modernisation paradigm by neopopulists. This paradigm, it was argued, involved 'top down' development, a narrow economic focus and a neglect of cultural and environmental aspects of development. The themes raised by the DHF have been developed and popularised in subsequent decades by neopopulists.

Potter *et al.* (1999) identify the growth of a community-oriented approach to development in the 1970s. There was a growing recognition that 'development from below' was needed, as those principally affected by development had little or no input into shaping it. A champion of neopopulist development strategies in the literature is Robert Chambers, who, in his writing, advocated 'bottom up planning', 'decentralisation' and 'participation' (Chambers 1983, 1988 and 1997). Terms such as these were to become characteristic of neopopulism, and have had a profound influence on debates about development. They are also characteristic of the advocacy of ecotourism ICDPs.

Chambers' approach was echoed in a growing body of work, and could, by the early 1990s, be considered to have become a significant and influential strand of thinking (Cooke and Kothari 2001). Stohr, for example, talks of organising from the 'bottom up' and, using the core periphery conceptualisation from dependency theory, from the 'periphery inwards' (Stohr 1981). In 1981 Stohr and Taylor described such an approach as 'development from below' (Stohr and Taylor 1981) and in 1983 Chambers termed this 'putting the last first' (Chambers 1983). Development had, it was held, to begin to 'put people first' (Cernea 1991) and, in order to achieve this, it was to be 'characterised by small-scale activities, improved technology, local control of resources, widespread economic and social participation and environmental conservation' (Ghai and Vivian 1992: 15). Friedmann (1992) argues that development in the context of developing world states should embrace self-sufficiency, self-determination and empowerment, as well as improving people's living standards.

Community participation is today a central part of development discourse (Potter *et al.* 1999: 9; UN 1993) and characteristic of the discourse on ecotourism examined in Chapters 4 to 7.

The anti-modern character of neopopulism

The rise of such community-oriented development could also be described as the rise of post-modern development (Potter *et al.* 1999: 13–14). This implies the rejection of unilinear development, or 'western developmentalism' as one neopopulist advocate has it (McMichael 2001: 34), and replaces it with a relativistic view of development that sees paths to development as essentially different in different circumstances.

Potter *et al.* (1999) point out that the growth of neopopulist development strategies has been accompanied by, and overlaps with, a discourse on development that began to reject modernity and the Enlightenment conception of progress. The critics rejected the meta-narrative implicit in modernity, seeing it as essentialist and dismissive of cultural differences. They rejected the universal side of modern conceptions of development – that urbanisation and ever greater ability to harness nature on a grand scale for human ends constitutes 'progress' – and in their place championed particular, local, strategies that are based on particular, local conditions (Potter *et al.* 1999: 8). Put another way, they rejected grand development projects in favour of micro projects – the sort of projects that today feature ecotourism as a form of integrated development and conservation.

From classes to people in neopopulism

We can also relate the growth of neopopulism to the decline of more traditional political channels for addressing development issues. It is generally accepted that traditional political identities, based around class divisions and focussed on the contestation of power at the level of the state, have declined in terms of their purchase on contemporary consciousness (Heartfield 2002; Gorz 1997; Touraine 1988). It is notable that in terms of social action, NGOs have grown in prominence as these more traditional channels for social action have declined in their ability to frame contemporary debates (Touraine 1998). The decline in traditional political identities and the lack of belief in, or commitment to, grand political ideas appears, by default, to elevate the role of civil society as an arena for human agency. Indeed, it has been argued that it is the failure of radical politics, in the form of neo-Marxism, to provide practical assistance to those on the front line of development that has turned erstwhile radicals towards locally oriented 'projects', and the world of NGOs (Amin 1985). As the efficacy of the state to resolve problems has declined, many have sought solutions in the realm of civil society, through NGOs.

Hence, neopopulism, a self-conscious reference to the importance of 'people', may in reality coincide with a narrowing of the scope for these same people to control their destinies both with regard to the developing world and the developed world too. Local participation affecting one's community is much more a part of contemporary political dialogue, yet significant choices for social change at a national level have declined. The extent to which neopopulism marks a significant and positive departure for communities in the developing world, with regard to ecotourism ICDPs, is a theme throughout the study. It is insightful to examine civil society further in this context.

Civil society

NGOs are 'civil society' organisations – they are neither part of the state nor are they commercial companies,[4] and hence fall into a broad category encapsulating human agency outside of the other two categories. The term civil society also has a close affinity with neopopulism. It is often associated with ideas of community, participation and reciprocity (Seligman 1992: 4).

This section looks at the idea of civil society, specifically in the way it informs the growth of NGOs and their practice in the realm of integrated conservation and development.

Origins of civil society

It is useful to look briefly at the derivation of the term to understand its salience in the contemporary politics of development. Civil society is a difficult term, partly because it has been defined in different ways by different people at different periods in history. Hegel was the first thinker to develop the concept, using the term *burgerliche gesellschaft*. Hegel's civil society has been succinctly described as a 'new sphere, [in which the] private and public, particular and universal, could meet through the interaction of private interests, on a terrain that was neither household nor state, but a mediation between the two' (Meiksins-Wood 1990: 62). Here, civil society lies outside of the state and the family, but mediates between the individual and the state. Hegel's conception holds something important for civil society today, as it identifies a realm in which individuals can be political actors outside of the institutions of government. Today, NGOs can be seen as playing this role.

However, the contemporary understanding of civil society has much to do with global developments associated with the end of the cold war. To a large extent, the term was popularised around the time of the collapse of the communist regimes in eastern Europe (Burgess 1997; Stompka 1992; Dahrendorf 1990). These regimes, it was held, dominated social life to the extent that a life outside of the state, embodying the subjectivity of 'the people', was restricted. Eastern Europe had witnessed 'the apparent atrophy or non-existence of the meso level of social relations, the sphere of social self organisation, and of that level in the articulation of interests that is to be found between the private realm of the domestic and the totalising state' (Marshall 1998: 74). Hence, the challenge facing post-communism became to reinvigorate civil society, including formal voluntary activity, such as that evident in NGOs. This perspective informed some of the interventions of former 'Western' countries to their 'Eastern' neighbours, such as initiatives to guarantee minority rights (Burgess 1997).

However, civil society has subsequently become increasingly pervasive in contemporary politics. For example, it is often employed to refer to the need to develop tolerance, lawful behaviour and institutions that will help to cohere societies in a diverse range of situations, from the rural developing world to Britain's cities. Indeed, one of the problems with the term is the breadth of its usage. For Kumar, '"Civil Society" sounds good; it has a good feel to it; it has the look of a fine old wine full of depth and complexity. Who could possibly object to it, not wish for its fulfilment?' (Kumar 1993: 377). Hence, the identification of NGOs with civil society has enabled them to acquire a certain moral status vis-à-vis governments and the commercial sector (Potter *et al.* 1999: 181). Further on this theme, Seligman comments on the moral status of civil society thus. It is 'a philosophically normative concept, that is – putting it in somewhat grandiose terms – an ethical ideal, a vision of the social order that is not only descriptive, but prescriptive, providing us with a vision of the good life' (Seligman 1992: 4).

Civil society can be described schematically and simply as that sphere of human subjectivity that lies outside of the jurisdiction of the state, and also is not a part of the commercial world. It hence reflects aspirations and concerns of what could be loosely described as 'the people'. NGOs are just one, but nonetheless probably the most important, expression of this amorphous civil society. Of course the term 'NGO' covers a wide variety of types of organisation, and this is reflected in this study. However, there is no doubt that the number, importance and influence of NGOs has increased since the 1980s, and that this is reflected in their increased role in tourism for integrated conservation and development in the developing world.

NGOs as exemplary of civil society

The above summary on civil society is pertinent to ecotourism ICDPs. First, the NGOs themselves are often considered to be exemplary of the increasing salience of civil society in politics, and specifically in development. Second, ecotourism ICDPs adopt a neopopulist approach to development, an approach that has a strong affinity with civil society – projects are seen as promoting civil society in the communities in which they operate.

To briefly develop the first of these themes, the NGOs featured in the case studies are themselves part of a broader growth of NGOs, and of the growth in importance attributed to 'civil society' in politics, and in development and conservation in particular (Princen and Finger 1994). Writing specifically about NGOs and tourism, Mowforth and Munt assert that '[t]he Socio-Environmental movement in its many guises has become one of the most enduring images of the last twenty years and has captured the public imagination in a way that has far surpassed other movements' (1998: 158).[5]

In similar vein, with reference to the rise of concern with the environmental effects of modern society, Eckersley comments that:

> [t]he environmental crisis and popular environmental concern have prompted a considerable transformation in western politics over the last three decades [. . .] [W]hatever the outcome of this realignment in western politics, the intractable nature of environmental problems will ensure that environmental politics (or what I shall refer to as ecopolitics) is here to stay.
>
> (1992: 7)

This transformation reflects, in part, the disillusionment with traditional political channels, and also with more traditional political ideas themselves. It has convincingly been argued that the grand political schemas of Left and Right have declined in their purchase on contemporary consciousness and that, in their place, we have witnessed the rise of a politics of the environment and of the community (Giddens 1995). This development is reflected in the growth in importance of NGOs and in the rise of neopopulist themes in development.

Civil society can also be presented as playing a distinctive role with regard to development and the environment in particular. In relation to development, the commercial sector is often viewed as exploitative of the developing world due to its emphasis on profit, and state-led development strategies have been heavily criticised on the basis that the aid is often tied to Western interests, or that the developing world state suffers from corruption (Potter *et al.* 1999: 181). In relation to conservation, business may be seen as neglecting negative environmental externalities and states again are often considered unresponsive to long term and non-economic impacts on the environment (Pepper 1996). Hence, NGOs have a certain moral authority in debates on sustainability through their counterposition to the governmental and commercial sectors (Edwards and Hulme 1992: 14), these latter two sectors apparently being tainted by the failures of the past.

Civil society and ecotourism

Not only are the NGOs exemplary of the growth of civil society, but civil society is also an important theme in ecotourism itself. As already mentioned, civil society has a strong association with neopopulism, an outlook that emphasises the role of non-state, non-commercial 'peoples' movements. For example, Watts points out that the growth in interest in civil society, taken to include 'communities, popular movements and social networks', provides 'the possibility of alternative (grassroots, participatory, subaltern) visions of development outside of the horizon of both state and market' (2000: 170). Just as the NGOs themselves are part of civil society, so too are 'communities' as often conceived – neither is part of the state nor directly motivated by commercial interest. In this vein NGOs may be well placed to promote community development (Brohman 1996a; Edwards and Hulme 1995 and 1992). NGOs are perceived as relatively free from commercial or political imperatives, and hence may be viewed as able to act outside of these constraints, in the interest of people and communities, in the most direct sense – as civil society organisations they are deemed to be in a good position to promote civil society themes in development. Hence, NGOs are not just a *manifestation* of civil society but, in the case of the case study organisations, they *invoke* civil society in their ecotourism projects. The role specifically of ecotourism ICDPs in promoting 'civil society' themes such as community and local identity is alluded to in Chapters 4, 5 and 7.

The literature on ecotourism

This section moves on to look at the specific literature on ecotourism. It situates the study in this literature, and shows how the former contributes something new to the latter.

There is a very large and varied literature on ecotourism that consistently emphasises its capacity to combine development and conservation (Fennell 2003; Scheyvens 2002; Goodwin 2000; Tourism Concern/Mann 2000; Wearing and Neil 1999; Neale 1998; USAID 1996; Budowski 1976). Also, a great deal of this literature carries a sense of *advocacy* for ecotourism as sustainable development in the rural developing world, comparing it favourably to mass tourism in this regard. It is no exaggeration to say that ecotourism has become, even in academic circles, casually equated with sustainable development (Butcher 2003a: 44).

Yet there is also a great deal of critical commentary and analysis on ecotourism in these and other books on the subject. This critical emphasis is a characteristic shared with this study. Many writers and activists who broadly advocate ecotourism are also intensely critical – they offer critical support to this new innovation in integrating conservation and development. For example, Pleumaron, in a paper entitled 'Ecotourism or Eco-Terrorism', argues that ecotourism 'can be just as damaging as honest, hedonistic holidaymaking' (Pleumaron 1995: 2). Ecotourism, unlike tourism to already developed regions, threatens 'the expropriation of "virgin" territories' (ibid.). Moreover, 'travellers have already opened up many new destinations' (ibid.), bringing the mass tourism that ecotourism's advocates seek to avoid. Cater, while arguing for ecotourism as potentially an environmentally and culturally benign form of rural development, sees similar dangers in the development of ecotourism: 'There is a real danger that ecotourism may merely replicate the economic, social and physical problems already associated with conventional tourism. The only difference [. . .] is that previously undeveloped areas are being brought into the locus of international tourism' (Cater 1992: 14).

Martha Honey, Director of the Ecotourism Program at the Institute for Policy Studies in Washington, DC, makes a similar point, asserting the following:

> By definition, ecotourism often involves seeking out the most pristine, uncharted and unpenetrated areas on Earth. Often, these are home to isolated and fragile civilisations. In some areas, eco-tourism is at the front line of foreign encroachment and can accelerate the pace of social and environmental degradation and lead to a new form of western penetration and domination of the last remaining 'untouched' parts of the world.
>
> (Honey 1999: 90)

Here, not only is ecotourism seen as complicit in destructive practices, but it is actually 'at the front line of foreign encroachment' (ibid.). Such pronouncements conjure up an image of cultural purity degraded by outsiders.

This fraught, self-critical advocacy of ecotourism is also well expressed by a writer in the American Audubon Society magazine:

> Tour boats dump garbage in the waters off Antarctica, shutterbugs harass wildlife in National Parks, hordes of us trample fragile areas. This frenzied activity threatens the viability of natural systems. At times we seem to be loving nature to death.
>
> (Berle 1990: 6)

Such comments are typical of the dilemmas within the advocacy of ecotourism.

This critical literature also includes frequent allusions to the sentiments of Wheeller (1993 and 1992) that ecotourism's claims are unrealistic, and simply act as a politically correct smokescreen covering the continued deleterious effects of a burgeoning tourism industry. Elsewhere, in similar vein, Butler (1992) has pointed out that many advocates of sustainable tourism associate it with smallness of scale and ecotourism, and hence fail to address the bulk of the very industry they seek to reform.

All these critics make trenchant points about the inconsistencies and problems inherent in the advocacy of ecotourism as sustainable development (see also Koch 1997; McIvor 1997). However, their criticisms often amount to support for the aims of ecotourism alongside the view that these aims may be difficult to achieve, or are naive with regard to the trajectory of the industry and the nature of a modern consumer society in which narrow self-interest (in this case the desire to travel) takes precedence over environmental ideals. The criticisms are debates *within the advocacy of ecotourism* and, as such, accept the premises of this innovation in thinking on development and conservation. This study, by contrast, attempts to critically examine these premises.

A good example of this limited critical outlook is a book entitled *Tourism for Development: Empowering Communities* (Scheyvens 2002). This takes as its starting point that development through tourism should mean the provision of a 'sustainable livelihood option to local communities' (ibid.: preface), and that this sustainable livelihood in rural areas should be built around the neopopulist themes of smallness of scale and local community participation, eschewing larger-scale developments (ibid.: preface and ch. 1). Once these themes have been asserted as constituting 'empowerment' and 'people centred development', the rest of the book looks critically at the extent to which these are, or could be, achieved in practice. The *validity of the aims themselves* is not considered.

That the aims of ecotourism – to link conservation and development together in rural areas, often in the developing world – constitute sustainable tourism development is largely supported in all the above analyses, or in the case of Wheeller (1993 and 1992) and Butler (1992), remains unchallenged. The advocates of ecotourism, though intensely critical of *practice*, all view the *aims* of ecotourism as a progressive step forward in relation to mass tourism, and as having potential for a more sustainable development in the rural developing world.

This lack of critical analysis of the premise of ecotourism led Munt to point out that, '[w]hile mass tourism has attracted trenchant criticism as a shallow and degrading experience for third world host nations and peoples, new tourism practices have been viewed benevolently and few critiques have emerged' (1994: 50). This is a telling comment – the premise of mass tourism, as an exemplar of modern development, has been subject to rigorous criticisms. Yet ecotourism, *as a development strategy*, has seldom been analysed. Indeed, in a sense the advocacy of ecotourism has acquired a certain moral status which shields it from criticism precisely through its counterposition to mass tourism, the latter assumed unethical (Butcher 2003a).

Of central importance here, then, is just such an analysis of ecotourism as a development strategy. The study seeks to critically analyse the assumptions underlying the association of ecotourism with sustainable development in the rural developing world.

Elsewhere, the contested nature of 'sustainability', with regard to tourism, has been the subject of book chapters and papers, most notably in Mowforth and Munt's *Tourism and Sustainability: New Tourism in the Third World* (1998). Indirectly, the present study also relates to work on the socially constructed character of 'nature' and 'culture', these concepts being central to sustainability (Urry and MacNaghten 1998; Urry 1996). If nature and culture are socially constructed, and hence subject to different assumptions, then sustainability must be too. To enquire into the character of ecotourism is to enquire into these assumptions, a perspective integral to this study.

The most thorough analysis of sustainable tourism, that looks beneath questions of definition to the social significance of the term, is Mowforth and Munt's text, referred to above. These authors provide a radical critique of sustainability and ecotourism, seeing them as concepts that in practice serve the purposes of various protagonists, most notably the commercial sector and those tourists who seek to differentiate themselves from the masses (as in 'mass' tourism). Their text draws on similar themes to this study, is interdisciplinary and is effectively an intervention in the debate rather than a commentary on it. However, while Mowforth and Munt put sustainable tourism development in a wider political and ideological context, they do not develop a critique of the main claim of ecotourism – that it can bring sustainable development on the basis of a symbiosis between conservation and development. Rather, their excellent critique focuses on pointing out the hypocrisy and inconsistencies in sustainable tourism in the developing world, when seen in context of broader power relations and inequality. In situating sustainable tourism, ecotourism and other forms of 'New Tourism' (Poon 1993) in wider power relations, Mowforth and Munt's (1998) critique tries to go well beyond those available elsewhere. This study shares this perspective and ambition.

Summary

This chapter has attempted to establish the development of the key ideas upon which ecotourism is premised. It has been argued that important strands of thinking on development, and on conservation, have tended to converge around neopopulism with regards to development in the rural developing world. It has also been argued that neopopulism is characterised by an emphasis on local community level development, development that involves participation by the community and encapsulates a symbiosis between this development and the conservation of the resources on which the community most immediately depend. This latter feature is often characterised as sustainable development.

Thus, neopopulist strategies have become more influential in part due to the growth of NGOs, themselves indicative of the growing salience of civil society in development and in politics more generally. Not only are civil society organisations, such as those featured in the case studies, more likely to be engaged in development in the developing world, due to the greening of aid, but they are also seen as being able to promote civil society themes such as community participation in the context of development.

The conceptual basis for ecotourism is characterised by neopopulism (with its emphasis on community), local level development and sustainability. The analysis attempts to establish a critique of this innovation in development practice. As outlined in the section on 'The literature on ecotourism' in this chapter, many critical commentaries on ecotourism are limited by their failure to engage with important premises and assumptions. This study attempts to open these up as areas for research.

Chapter 3 considers the sample case study organisations' aims, the roots of their involvement with ecotourism as sustainable development in the rural developing world, and identifies a shared rationale for this. This provides a bridge to a critical examination of the neopopulist premises of the advocacy of ecotourism in the succeeding three chapters.

3 Pioneers of ecotourism

Different aims, shared perspective

Introduction

This chapter looks at the aims of key NGOs involved in pioneering and refining ecotourism as a strategy to integrate conservation and development. Their view of sustainable tourism development as applied to the rural developing world is established and ecotourism is considered in this context.

A different approach is taken with regard to the UN International Year of Ecotourism, as this is a conference featuring NGOs rather than a specific organisation. The chapter discusses the IYE in the context of a series of keynote conferences and reports in order to establish its antecedents and trajectory.

Finally, and strikingly, it is apparent that organisations with different formally stated aims – broadly conservation and development/well-being respectively – coalesce around a common rationale for ecotourism. This rationale is, put simply, that ecotourism has the potential to generate a 'symbiosis' between these divergent aims.

We begin with the largest and most influential international player in conservation, the World Wide Fund for Nature.

The case studies and their relationship to ecotourism

WWF

WWF is the largest global conservation NGO in the world. It operates on the basis of fifty-two offices, consisting mainly of country offices, but also some based on regions and project offices, as well as its international operation, WWF-International. It is active in conservation in over ninety countries (WWF-UK undated c). It claims five million supporters worldwide. Some 90 per cent of its resources come from 'voluntary' sources, comprising commercial and private donations (ibid.).

WWF was launched in 1961, originally in the UK, but spreading rapidly around the world. In the 1970s its emphasis broadened from wildlife conservation to looking at the effects of human activities on the environment.

In the 1980s, WWF was instrumental in the development of the influential *World Conservation Strategy*, which linked conservation explicitly to humanity's future – humanity, the strategy held, had to learn to live within pressing environmental limits were it to have a future (IUCN *et al.* 1980). The production of this important document in conjunction with two UN bodies, the IUCN and UNEP, is also exemplary of the close links that WWF has maintained with global institutions such as the UN. For example, at the UN Johannesburg World Summit on Sustainable Development in 2002, WWF was the only conservation organisation invited to address the heads of state present.

WWF argues that conservation has to be seen alongside the needs of communities. It is, according to its publicity, 'For People and For Nature':

> WWF has long believed that you cannot eliminate poverty without protecting the environment and you cannot protect the environment without tackling poverty – the two issues are inextricably linked. We therefore spend half our conservation funds on promoting *alternative livelihoods* and helping people manage their environments. By promoting sustainable development throughout the world, WWF aims to ensure that the benefits and wonders of nature remain to be enjoyed by future generations.
>
> (WWF-UK undated a; my italics)

Ecotourism is just such an alternative livelihood, often promoted as an alternative to other options deemed destructive of the natural environment.

WWF's stated mission is 'to stop the degradation of the planet's natural environment and to build a future in which humans live in harmony with nature', by 'conserving the world's biological diversity'; 'ensuring that the use of renewable resources is sustainable'; and by 'promoting the reduction of pollution and wasteful consumption' (WWF-International 2001b: 1). This represents the conservation imperative that underlies all WWF's work. The notion of 'harmony' is important here, and is alluded to in Chapter 7 and elsewhere in this study.

WWF seeks to 'reconcile human development needs with those of biodiversity conservation within large-scale areas' with the aim of 'ecoregion conservation' (WWF-International 2001a: 3). The ecoregions consist of 240 regions around the world that are unique in terms of their biodiversity. WWF seeks to influence the way 'natural resources and the environment are used and changed by people' in these regions (ibid.). Further, '[w]here tourism is a major activity in an ecoregion, it is important that the conservation vision and strategy for that ecoregion takes account of the threats and opportunities posed by tourism' (ibid.).

In addition, 'tourism should be integrated into broader regional priorities' (ibid.). With regard to ecotourism in the rural developing world, these priorities revolve around biodiversity conservation. It is worth quoting WWF documentation at some length to establish its relationship to tourism development in the developing world:

> In certain areas that are particularly ecologically fragile, any form of tourism development may be inappropriate. Tourism is more acceptable, however, where its potential negative impact is judged to be less than that which might result from alternative development strategies such as mining or logging, or where the development of part of an area for tourism allows the remainder to be conserved.
>
> (WWF-International 2001a: 3)

Indicative of the growing importance of tourism in the 1990s was the appointment of Justin Woolford as Tourism Policy Officer in the International Policy Unit at WWF-UK, a unit that also works closely with WWF-International on issues such as biodiversity conservation, the Convention on Illegal Trafficking in Endangered Species (CITES) and climate change (Woolford 2002). Other NGOs, including CI and SNV, created new posts in this area around the same time.

WWF's approach to tourism development, and indeed all development, is 'precautionary' (WWF-International 2001a: 1; WWF-International 2001b: 2).[1] There is also, however, pragmatism, given the recognition of the importance of community well-being. This is especially pertinent given the criticisms made of WWF's conservation activities in the past as being akin to 'fortress conservation' (e.g. see Mowforth and Munt 1998: 117 and 167). In this spirit WWF accepts that, '[t]ourism may be acceptable, even in the most sensitive environments, if it contributes to *sustainable livelihoods within a community*' (WWF-International 2001a: 3; my italics), referring to livelihoods that maintain, and do not transform, the communities' relationship to their natural resources.

WWF also has extensive links to government and supragovernmental bodies. A notable example is the *World Conservation Strategy*, produced jointly between the UN and WWF. It remains an oft-cited keynote document in the development of sustainable development in global government. WWF participates in the Commission for Sustainable Development (CSD) and the UN Biodiversity Convention, which has a bearing on its attitude towards tourism, as does its links to the UNEP and UNESCO (Woolford 2002).

Woolford believes that there are 'a lot more similarities than differences' between the various NGOs concerned with looking at tourism (Woolford 2002). Although WWF 'fall[s] one side of the environment/development debate, or even conflict' (ibid.), it is keen to work with all the different organisations. Woolford simply regards it as constructive to take this approach, as for WWF, 'there isn't really a conflict between environment and development' (ibid.). Also, while he accepts that there are smaller, vocal NGOs who regard WWF and CI as being 'corporate', he argues that there is 'strength in complementarity' (ibid.) between those more critical of business and those looking to work with business.

While WWF's engagement with the tourism industry is broad and varied, it is underwritten by its principal aim of environmental conservation. Ecotourism ICDPs are one important aspect of this. The coincidence of ecoregions with rural poverty has contributed to ecotourism becoming an important innovation in their work, as ICDPs claim to address both priorities simultaneously.

CI

CI is a US-based conservation NGO. Its headquarters is in Washington, DC, and its work covers more than thirty countries in four continents. CI was founded in 1987, in part based on a critique of existing conservation practice, as an alternative to what it saw as an exclusionary 'fortress conservation' approach (Sweeting 2002). A split emerged in Nature Conservancy,[2] a leading American conservation NGO, and from this CI established itself as an advocate of conservation that emphasises the needs of local communities. This is effectively the philosophy of community conservation, and community-based ecotourism fits well with CI's aims.

Its stated mission is 'to conserve the Earth's living natural heritage, our global biodiversity, and to demonstrate that human societies are able to live harmoniously with nature' (CI undated a). To achieve this, it 'appl[ies] innovations in science, economics, policy and community participation to protect the Earth's richest regions of plant and animal diversity in the hotspots, major tropical wilderness areas and key marine ecosystems' (ibid.).

The biography of Jamie Sweeting, Director of the Travel and Leisure Industry Initiative in the Center for Environmental Leadership in Business (CELB), is evidence of the growing salience of tourism to these wider aims. Sweeting joined CI in 1995. Prior to this he worked as an information specialist for The Ecotourism Society,[3] and conducted a specialist management project on ecotourism in the Caribbean (Sweeting 2002). He has also managed the Ecotourism Programme in CI's Conservation Enterprise Department. The work there involved close collaboration with CI's field offices to develop and implement ecotourism projects in more than a dozen countries. Among his achievements, he oversaw the development and launch of the Ecotravel Center, a comprehensive ecotourism internet resource linking the projects to the market through niche tour operators (ibid.).

CI looks to utilise small-scale nature-based tourism to generate integrated conservation and development. However, it also sees more mainstream tourism as a danger to biodiversity. In fact, it is mentioned as such in relation to the majority of biodiversity hotspots, these being the places it prioritises in its work based on the rich biodiversity within them (e.g. CI undated b). Almost all of these hotspots are in less developed or middle-income countries. Here, ecotourism is a tool in CI's armoury for achieving conservation and development.

CI, along with the WWF, is an international NGO with high levels of corporate funding. It engages with tourism at all levels of the industry. However, in the developing world, in CI's priority 'hotspots', community-based ecotourism has emerged as an important strategy for promoting what they view as sustainable tourism.

SNV

SNV describes itself as a 'multicultural development agency' (SNV undated a), stressing its allegiance to a notion of development that takes in cultural as well as economic dimensions. It is based in the Netherlands, but operates internationally. It supports organisations in twenty-eight countries in Africa, Asia, Latin America and Europe – often local NGOs or local government organisations.

Its stated mission and core business are succinctly defined as, '[c]apacity building support to meso-level organisations and local capacity builders in their relation to structural poverty alleviation and improved governance' (ibid.)[4]. It seeks to improve the performance and increase the influence of the organisations it works with and, in so doing, promote the delivery of services to groups of poor people in remote areas. Private sector development in rural communities is also an important issue for them (ibid.).

Elsewhere, SNV's mission is described as being 'to develop and share knowledge and skills with local organisations with the aim of better equipping them for their work in structurally alleviating the poverty of both men and women' (SNV/Caalders and Cottrell 2001: 4). The statement reflects its emphasis on tackling rural poverty, and also on addressing gender inequality – clearly development or well-being goals, as opposed to the conservation orientation of CI and WWF.

Specifically in relation to tourism, SNV defines sustainable development as:

> [a] balanced target group oriented development strategy involving: socio-economic development and economic empowerment; local participation; social and political empowerment; economic sustainability; ecological sustainability; socio-cultural consciousness; and improving gender equality.
>
> (SNV/Caalders and Cottrell 2001: 4)

Hence, tourism is viewed as a multidimensional tool for achieving some of SNV's aims, principally in relation to community well-being, community empowerment and also conservation.

SNV originated in 1963 as a small organisation posting Dutch volunteers to the developing world. By the start of the 1980s there was a realisation in the organisation that it could play a role in raising awareness in the Netherlands of issues pertaining to the developing world. Also, rather than being a voluntary organisation, it was increasingly sending well-paid experts to work overseas to help in achieving its development goals. The acronym SNV has been retained, even though it does not reflect the organisation's activities. It has been added to, though, with the full title of the organisation now being SNV Nederlandse Ontwikkelingsorganisatie, or SNV Netherlands Development Organisation (ibid.). However, the organisation is generally referred to by the acronym SNV.

From the 1980s onwards, the emphasis shifted towards technical assistance and advice. The advice is aimed at enabling local people in the countries in question to develop their own initiatives in line with SNV's aims (SNV 2000: 7–8). Often this is discussed in terms of involving other 'stakeholders' and 'empowering' rural communities to take control of their own development. This is very much in the neopopulist spirit of 'endogenous development', as referred to in Chapter 2.

SNV's status during these developments is described in its literature as a semi non-governmental organisation. It continues to receive funding from the Dutch Ministry for Development Cooperation, but since 1991 new articles of association have stated that the SNV board could formulate and implement policy independently (SNV undated b).

In 1996 SNV specifically reformulated its commitments – its core business was now explicitly technical assistance. Four 'product groups' were defined:

- capacity building;
- project implementation;
- mediation;
- service provision for northern organisations.

(SNV 2000)

Programmes were geographically concentrated in marginal areas within the countries in which SNV operated. In general this refers to rural areas in the developing world – areas in which ecotourism ICDPs have emerged as a development tool. Also, more emphasis was to be put on SNV's 'mediating role between the different development actors' (government, NGO and private sectors), and on linking local organisations to actors at other levels within their countries (ibid.: 7).

Finally, and significantly with regard to this study, an interdepartmental policy evaluation in 1999 concluded that the quango status of SNV should be terminated, and this has led to a severing of SNV from the formal institutions of the Dutch government (ibid.: 9). Rather, like typical NGOs, the relationship had become and remains one of subsidiser and subsidy recipient.

With specific regard to tourism, it is unsurprising that ecotourism should be an issue for SNV, given its commitment to marginal, rural parts of the developing world, and its desire to link up with the commercial sector in its projects. This interest was formalised by the appointment of Marcel Liejzer in 1994. Liejzer had been involved in a project with Franz de Man of the Retour Foundation, a small NGO promoting sustainable tourism initiatives. SNV gave him the assignment of exploring the possibility of developing a project involving the Masai in Tanzania, work not dissimilar to that which he had carried out with the Retour Foundation (Leijzer 2002).

At this time SNV was 'one of the first development organisations to start advising communities on developing tourism projects. At that time it was seen as an almost revolutionary step' (SNV/Caalders and Cottrell 2001: 3). By the beginning of the twenty-first century, SNV had either finalised work, or had tourism ICDPs in progress, in Albania, Bolivia, Botswana, Benin, Cameroon, Ecuador, Ghana, Nepal, Niger, Peru, Tanzania, Uganda and Vietnam (ibid.).

The project with the Masai was deemed a success and, at the start of 1995, Leijzer became the first SNV Tourism Officer. The work in Tanzania developed into the Cultural Tourism Programme there, the experience from which was recorded in order to inform future practice (SNV/de Jong 1999). For example, in 1999 Leijzer looked at the possibilities for a similar project in Bolivia, although on this occasion it was decided that it was not possible to develop the project further.

Notably, Leijzer's role changed as of 2000, when he became Private Sector Officer (Leijzer 2002). Tourism still lies within his remit, as it is considered vital that projects relating to tourism link up with the private sector in order to be self-sustaining in terms of revenue and development potential. Indeed, tourism has, according to Leijzer, grown in importance for the organisation (ibid.). Agriculture has been, and continues to be, its main focus in the rural areas it prioritises, but tourism has emerged as an alternative or complement to agriculture in such areas (SNV/Caalders and Cottrell 2001).

SNV works closely with the private sector, including its work on tourism. It also maintains close links to the Dutch government through various channels, some of which are pertinent to tourism. For example, through its links with the Dutch government's Centre for Promotion of Imports (CBI) it supports small- and medium-sized businesses in reaching the market in Holland. This includes the crucial area of marketing support. An example of this relates to community tourism in Nepal, where SNV Nepal has linked up with the CBI to develop training in ecotourism marketing for the Nepalese (SNV/Caalders and Cottrell 2001: 15).

However, SNV also works extensively with non-Dutch international development agencies in tourism related projects. The organisations include USAID, UNDP (UN), IUCN, Finida (Finland), GTZ (Germany), DfID (UK) (SNV/Caalders and Cottrell 2001: 17). For example, USAID funds a project proposed by SNV-Ghana and the Ghana Tourist Board and, in Tanzania, SNV worked with GTZ and Finida. SNV has worked in Botswana on ecotourism ICDPs funded by the UK government (SNV/Rozenmeijer 2001). The Asian Development Bank and the UNDP have also been partners (ibid.).

So SNV, from a rural development perspective, has pioneered ecotourism to meet its aims. It is deemed to be important in key contemporary development priorities – establishing basic needs and poverty reduction – as well as being regarded as having the potential to constitute sustainable development.

Tourism Concern

Tourism Concern was founded by Alison Standcliffe in 1989, originally being run from her home in Newcastle. Standcliffe had been working in Singapore, and became conscious of how poverty provided a photogenic opportunity for tourists and how commercial considerations distorted the relationships between tourists and their hosts (Barnett 2000). She was introduced to the Tourism European Network (TEN), which at the time was dominated by the Ecumenical Coalition on Third World Tourism (ECTWT) (Barnett 2000; Botterill 1991). Today, Tourism Concern is prominent within TEN, which remains as an umbrella for similar organisations elsewhere in Europe (Barnett 2000). TEN is distinctive because it consists of membership organisations.

The events leading to the establishment of Tourism Concern have been recorded by one of their original steering group members, David Botterill (1991). In 1988, a meeting was organised, drawing on the mailing list of the ECTWT to publicise it (Botterill 1991). The meeting provided the impetus to formalise a steering group, and the decision was taken to turn the group into a membership organisation in July 1989 (Botterill 1991: 205) A steering committee was formed comprised mainly of academics, and Tourism Concern was launched (Barnett 2000; Botterill 1991).

Botterill argues that these humble beginnings were of some moment – that this was part of 'the emergence of a new global social movement' (Botterill 1991: 203). Certainly, the growth in influence of Tourism Concern, alongside the more general critical focus on tourism in academic writing, emanating from journalism, NGOs and campaigns (Butcher 2003a: ch. 1), would suggest that there is some truth in this rather grand claim.

In the early 1990s Patricia Barnett was employed as their single worker. By June 2000, Tourism Concern had eight workers, four full-time with the others part-time, some working from home (Barnett 2000). From its birth, the group quickly established some 200 members, many of whom work in the higher education sector, and this rose to around 1,000 by 1995 (Mowforth and Munt 1998: 159). Subsequently, this number has remained fairly stable. However, as Barnett argues, it is the success of the campaign in getting their point across that has shown the most impressive growth (Barnett 2000). This is highly evident to anyone who has been following the debates over a number of years. One notable event in this regard was a boycott campaign in 2000 aimed at *Lonely Planet Guides* which refused to withdraw its guide to Burma, a regime that has used repressive methods and coerced labour in the course of developing its tourism infrastructure (Tourism Concern undated d). The subsequent campaign gained Tourism Concern unprecedented publicity in the national media.

Tourism Concern was founded due to a concern about the effects and trajectory of commercial tourism. Its initial aims included promoting 'greater understanding of the impact of tourism on host communities and environments' (Botterill 1991: 207), reflecting the view that global tourism had all too often ridden roughshod over environmental and social issues pertaining to host societies. Tourism Concern has tended to be critical of mass tourism, and has promoted small-scale tourism, especially 'community tourism'. However, while it is often critical of what it views as the excesses of the industry, past and present, it has also cast a critical eye onto the newer, nature-based niche markets such as ecotourism. While it advocates community-based ecotourism as sustainable tourism development, and also as 'fair trade' (Tourism Concern 2000), it remains wary that 'green washing' can be employed by companies to present an ethical front to potential customers and that, in spite of the rhetoric of 'community', conservation organisations remain prone to prioritise conservation above community well-being (Barnett 2000).

Tourism Concern has also made attempts to influence the mainstream tourism industry. For example, in 1999 it held a meeting attended by seventeen out of twenty-five invitees from the industry (Barnett 2000). Along with Voluntary Service Overseas (VSO) it has produced in-flight videos to promote ethical holidaymaking for package tourists. One of these is on First Choice flights going to the Gambia, and an edited version is used by the Gambia Hotels Association. Similar videos relating to Kenya, Taiwan and the Caribbean have been produced, in conjunction with VSO (Barnett 2000). These are good examples of Tourism Concern working with the industry in order to educate tourists to promote 'ethical' consumption in tourism, and this is linked to tourism's prospective development impact.

Tourism Concern clearly wants to influence the mainstream industry. Indeed, it is reported to have redefined its mission statement to reflect this priority (Lara Marsh of Tourism Concern, cited in Scheyvens 2002: 186). However, the types of tourism promoted by Tourism Concern as sustainable tourism development are out of step with the majority in an industry that consists of large, multinational profit-maximising companies. Tourism Concern's web site is replete with links to organisations promoting small-scale, community-based ecotourism, and its best selling publication, *The Community Tourism Guide* (Tourism Concern/Mann 2000), sets out a vision of sustainable tourism in the developing world that is simply incompatible with the modus operandi of the bulk of the industry. The distance between Tourism Concern and the industry is great on the question of sustainable tourism in the developing world. Tourism Concern clearly favours a sea change in consumption and in the industry towards small-scale community-based tourism, but this would undermine the existing trade of many tour operators.

One stark illustration of this is the case of the Gambia. In the Gambia, Tourism Concern, along with sister organisation Gambia Tourism Concern, lobbied hard against the development of all-inclusive resorts on the basis that they did not provide economic links to the local economies in which they were based. A subsequent ban on the development of all-inclusives by the Gambian government was criticised by big names in the industry on the basis that this would hit visitor numbers and hence the economy. The ban was later removed (Tourism Concern undated b).

Tourism Concern's relationship to the industry appears to be double-edged. On one hand, it clearly challenges the industry on a number of fronts without fear or favour (it may be added that its reliance on membership rather than on corporate funding supports a freedom to be critical). Yet, on the other hand, the industry cannot dismiss it, as it clearly reflects concerns held more broadly within society about modern tourism's impact on host societies (Mowforth and Munt 1998: Ch. 6).

The picture is not dissimilar with regard to its relationship with governmental institutions. It is notable that Tourism Concern's very first campaign, in 1991–2, was aimed at getting tourism onto the agenda at the UN Conference on Environment and Development in Rio. *Beyond the Green Horizon: Principles of Sustainable Tourism* (Eber 1992) was produced, jointly with the WWF, to push for this (Barnett 2000). Barnett points out that it took seven years for this to come to pass – tourism is now part of the biannual Commission for Sustainable Development, which is effectively an update on developments from the Rio conference (ibid.). Barnett herself has represented European NGOs at the CSD (ibid.).

Barnett believes that Tourism Concern's status as a critical, campaigning lobby group means that it is 'held in suspicion' by some organisations in and out of government concerned with similar issues (Barnett 2000). However, although its campaigning stance may make it an awkward partner at times for government agencies, it has been successful in developing channels of communication with government. For example, in the late 1990s it held a meeting, backed by the Department of Trade and Industry, with all relevant government departments (Department for International Development, Department of Environment, Transport and the Regions, Department of Culture, Media and Sport, and the Foreign and Commonwealth Office), to talk about Tourism Concern's human rights and other work. However, for Barnett, the departments remain 'huge fortresses' (ibid.) and, with the rapid change of personnel, consultation becomes difficult.

DfID did, however, consult Tourism Concern over the department's Tourism Challenge Fund, a fund initiated by a government White Paper in 1997 (DfID undated) to promote 'Pro-poor tourism' in line with the department's general perspective to aid 'the poorest of the poor' in the developing world (Ashley *et al.* 2000: 1).[5] DfID also provided some funding over three years (running from 1999 to 2002) for Tourism Concern's fair trade work (Tourism Concern undated a; Tourism Concern 2000). However, Tourism Concern lacks core funding from government bodies – its funding links are in areas where there is a limited convergence of interest over specific projects (Barnett 2000).

Botterill's (1991) paper, *A New Social Movement? Tourism Concern – the First Ten Years*, is unusual in that the author is writing about the development of an organisation within which he himself was a central figure – Botterill positions the emergence of what was at the time a tiny emergent campaign in the early 1990s in the context of new social movement (NSM) theory. He cites approvingly Touraine, who argues that NSMs are indicative of 'a struggle for recognition and control within [the social sphere]' (cited in Scott 1990: 62). The phrase 'struggle for recognition' is interesting here. Other social theorists have examined the idea of recognition as a growing focus for social action. Axel Honneth, developing the sociology of German theorist Jurgen Habermas, sees the struggle for recognition as a central component of modern social struggles (Honneth 1996). It implies that new and wider voices are being brought into societal discourse, that new points of view are being recognised. For Touraine, such NSMs are positive developments representing an emergent post-class political subjectivity (Touraine 1988, see also Scott 1990). Touraine's argument sees the collapse of more traditional class based social movements – those with a distinctive constituency pertaining to the realm of production – as opening the way for 'new' movements, new in the sense that they do not take social class as their primary point of reference. For Touraine, we live in a post-industrial society in which such traditional class based social movements are less relevant (ibid.), and for Botterill, Tourism Concern is part of this new subjectivity.

It is insightful that Tourism Concern has maintained a strong interest in promoting ethical consumption and ethical lifestyles. NSMs, operating in an environment in which traditional politics fails to inspire, have an affinity with cultural politics and the politics of consumption. For example, Tourism Concern has produced *The Community Tourism Guide*, which sets out small-scale, community-based holidays as being ethical for developing world societies on the basis of their economic, environmental and socio-cultural impacts (or in the latter two cases, lack of them). Also, a substantial part of its work is aimed at the consumer, with, for example, a number of leaflets having been published with titles such as 'Be different on your holiday' and 'When you travel, do you get concerned?'. It launched a code of conduct for young travellers in conjunction with *Rough Guides*, listing 'dos and don'ts' for ethical holidaymaking (Tourism Concern undated c).

Although the rubric 'social movement' is debatable, Tourism Concern certainly reflects a widely held view that mass tourism has proved highly environmentally and culturally problematic. This view is manifested in similar campaigns throughout the developed world – it is, for example, part of the Tourism European Network. Among such campaigns small-scale community-based ecotourism is advocated as a progressive alternative to mass tourism and other large-scale development, on the basis that it is relatively environmentally benign and culturally appropriate, a position similar to WWF, CI and SNV.

The UN IYE

The process of the UN IYE

In 2000 the UN declared that 2002 was to be the International Year of Ecotourism. The purpose of the event was:

> to bring together governments, international agencies, NGOs, tourism enterprises, representatives of local and indigenous communities, academic institutions and individuals with an interest in ecotourism, and enable them to learn from each other and identify some agreed principles and priorities for the future development and management of ecotourism.
>
> (UNEP/WTO 2002a: 7)

From this, it was hoped to achieve 'the setting of a preliminary agenda and a set of recommendations for the development of ecotourism activities in the context of sustainable development' (UNEP/WTO 2002b: 65).

The process leading to the Quebec Summit, from which emerged the *Quebec Declaration on Ecotourism* (UNEP/WTO 2002b) and *The World Ecotourism Summit Final Report* (UNEP/WTO 2002a), was a thorough and extensive process of discussion involving many stakeholders.

A number of regional panels were convened, raising general issues, as well as those pertaining to their respective regions. The four regional panels were as follows:

1. Maputo, Mozambique, March 2001 – for all African states with an emphasis on planning and management;
2. Nairobi, Kenya, March 2002 – for East Africa;
3. Mahe, Seychelles, December 2001 – for Small Island Developing States (SIDS) and other small islands;
4. Algiers, Algeria, January 2002 – for Desert Areas.

(UNEP/WTO 2002a: 11, 14, 16 and 18)

The regional panels were each to consider four themes, these themes themselves the result of prior consultation with various stakeholders. The themes were:

- Ecotourism Policy and Planning – the Sustainability Challenge;
- Regulation of Ecotourism – Institutional Responses and Frameworks;
- Product Development, Marketing and Promotion of Ecotourism – Fostering Sustainable Products and Consumers;
- Monitoring Costs and Benefits of Ecotourism – Ensuring Equitable Distribution Among All Stakeholders.

(ibid.: 7–8)

Drawing on the discussion at these panels, four main papers were written, one on each of the four themes, and these fed into the *Final Report* and the *Quebec Declaration* (ibid.: 8).

Overall, eighteen preparatory meetings were held in 2001 and 2002, under the aegis of either the WTO or the UNEP (in association with The International Ecotourism Society), involving over 3,000 representatives from national and local governments, private ecotourism businesses and their trade associations, NGOs, academic institutions and consultants, inter-governmental organisations and indigenous and local communities.

The World Ecotourism Summit itself was held in Quebec City, Canada, between 19 and 22 May 2002. It was attended by 1,169 delegates from 132 countries. The range of delegates included international conservation NGOs, development NGOs, national ministries of tourism, culture and the environment, private sector enterprises involved in ecotourism and finally academics and consultants (ibid.: 8–9).

The summit began with a plenary session at which reports from the four regional panels were presented. On the agenda were four parallel working group sessions covering the four main summit themes, a ministerial forum, two special forums covering the business perspective and development cooperation in ecotourism respectively, and a further plenary session to receive and debate reports from the four thematic working groups. The summit ended with a plenary session to receive and debate the draft *Quebec Declaration on Ecotourism*.

Also, alongside this formal process, a web conference was organised during April 2002, again involving many individuals from around the world. This important part of the process enabled any individual or body, at no cost, to contribute to the process. The discussions on this web conference were also to be considered in the production of the *Final Report*.

The process produced two documents that comprise the principal documents of this case study: the lengthy *World Ecotourism Summit Final Report* (UNEP/WTO 2002a); and the brief *Quebec Declaration on Ecotourism* (UNEP/WTO 2002b). The latter is included within the former, comprising pages 65–73, but is widely available separately and hence appears as a separate reference in this study. The *Quebec Declaration* sets out a list of principles and recommendations to be disseminated to governments, the private sector, NGOs, community-based associations, academic and research institutions, inter-governmental organisations, development agencies, financial institutions and indigenous and local communities (ibid.: 3). The *Final Report* includes reports from the regional panels, from the four thematic working groups and from the special forums on development cooperation for ecotourism and the ecotourism business perspective respectively.

Both documents formalise an existing trend for ecotourism to be advocated as sustainable tourism development, and also promote this for the future. As such, they are important documents pertinent to this study.

The process in context

The IYE's association of ecotourism with sustainable development in rural areas is striking. *The Quebec Declaration on Ecotourism* asserts that ecotourism 'embrace(s) the principles of sustainable development' (UNEP/WTO 2002b: 1). Indeed, the basis for the advocacy of ecotourism is that it is exemplary sustainable development, and that, as such, it can 'provide a leadership role' to the rest of the industry (ibid.). The *Declaration* instructed the World Summit on Sustainable Development (WSSD), the global summit on environment and development, held in Johannesburg in 2002, to 'recognise the need to apply the principles of sustainable development to tourism, and the exemplary role of ecotourism in generating economic, social and environmental benefits' (ibid.: 7). It is hence instructive to consider briefly the UN IYE in the context of the development of sustainable development through the UN.

The rise of sustainable development can be charted through the discussions, conferences and statements of the UN, which is ostensibly the highest level of global governance. The UN Conference on the Human Environment held in Stockholm in 1972 is sometimes cited as an important watershed in establishing environmental conservation in the face of rapid post-war economic growth (Adams 2001: Ch. 3). It was the first international conference specifically addressing perceived environmental destruction resulting from development, and it established the UN Environment Programme, which has played the leading role in developing and promoting debate and practice on sustainable development. Indeed, the UNEP was the prime mover, alongside the World Tourism Organisation (WTO), of the IYE itself.

More commonly cited in the rise of sustainable development are the *World Conservation Strategy* of 1980, compiled by the International Union for the Conservation of Nature in conjunction with the UNEP and the World Wildlife Fund (former name of the World Wide Fund for Nature), and also the UN commissioned report *Our Common Future*, often referred to as the *Brundtland Report*, of 1987 (Adams 2001: Ch. 3; Hall 1998).

The 1992 UN Conference on Environment and Development marked the most important watershed for sustainable development, placing the view that development should have a greater emphasis on its environmental and socio-cultural effects at the centre of political debate (UN 1993). It spawned discussions about sustainable development as applied to the tourism industry – i.e. sustainable tourism development. It is from this source that the International Year of Ecotourism emerged, through the UN, as a championing of ecotourism as exemplary sustainable development in rural areas, most often in the developing world.

The most commonly cited formulation of sustainable development, as developed through these UN conferences, is based on the principle of intergenerational equity. Sustainable development was to be 'development that meets the needs of the present without compromising the ability of future generations to meet their own needs' (UNEP 1993; WCED 1987: 43). This underpinning definition of sustainable development was established in *Our Common Future* in 1987, and popularised at the UN Conference on Environment and Development in 1992. A second important formulation is the oft referred to 'triple bottom line' that implicitly criticises the narrow economic focus of 'development' as previously constituted by placing environment and culture alongside economy as priorities for the new 'sustainable development' (Muller 1994). Both views inform the subsequent advocacy of ecotourism as sustainable development evident in the IYE and elsewhere.

What the above formulations represent is an attempt to bring a greater synthesis between development and the environment, priorities often seen as being in conflict with one another. However, in each case, attempts to formalise sustainable development as a new orthodoxy have been characterised by disagreements, with some claiming it restricts development and others arguing that conservation does not have a high enough priority (Adams 2001; Redclift 1990). Indeed, it is possible to argue that the UN conferences have been a focus for the tension between development needs and conservation as much as they have resolved this tension (ibid.). The IYE is very much a part of this lineage, and, it will be argued, the aforementioned tension is resolved there in favour of conservation and against development.

Summary

A number of important points emerge from the profiles of the organisations. All five of the case studies exhibit a desire on the part of the NGOs to influence the industry as a whole, which often means working with large-scale global tour operators. However, the forms of tourism considered sustainable in the context of the rural developing world tend to eschew global business, and development of any great scale, in favour of small-scale, community-oriented ecotourism, which, despite rapid growth, remains very much a niche market.

For CI and WWF, their core biodiversity conservation work focuses on the developing world, as it is here that the large majority of CI's 'biodiversity hotspots' and WWF's 'ecoregions' are located. These countries are characterised by their lack of economic development and, moreover, the regions within these countries prioritised for conservation are typically among the poorest. The two organisations, however, place great emphasis on the community benefits that can accrue *through* conservation, and argue that ecotourism can therefore constitute sustainable development.

SNV is interested in promoting sustainability widely within the tourism industry. However, its principal aim is development in rural areas in the developing world. It has also adopted community-based ecotourism as a tool for its work, arguing, along similar lines to CI and WWF, that it can constitute sustainable development.

Tourism Concern has sought to influence the direction of the tourism industry in a number of ways through their campaigns and initiatives. However, as *The Community Tourism Guide* and their fair trade work make clear, community-based ecotourism is deemed to have the potential to establish exemplary sustainable development in the rural developing world. Although Tourism Concern and SNV remain critical of what they see as the 'conservation first' emphasis of some organisations, this is essentially a disagreement *within* the advocacy of ecotourism rather than a rejection of the strategy.

Finally, the UN IYE marks a watershed in the growth in advocacy of ecotourism as exemplary sustainable development in the rural developing world. NGOs were centrally involved in the UN IYE, alongside governmental, industry and individual contributions. *The World Ecotourism Summit Final Report*, and accompanying *Quebec Declaration on Ecotourism*, comprise a key case of the contemporary advocacy of tourism ICDPs (UNEP/WTO 2002a and 2002b).

So, despite the *differences* between these case studies – two approaching the issue from a conservation position, two from a development/well-being one, and one a broad attempt to codify experience and best practice involving both perspectives – it is evident that ecotourism emerged in the 1990s as a *common* tool for integrating conservation and development in the rural developing world.

Moreover, a common rationale is evident too, and it is this that is interrogated in succeeding chapters. This rationale holds that the promotion of sustainable tourism in the rural developing world involves development based around conservation, rather than compromising it. In order to qualify as sustainable development, it should be small-scale, conserving local natural capital, and drawing on traditional knowledge and skills. This theme has been developed widely in the literature on ecotourism (Fennel 2003; Goodwin 2000; Honey 1999; Wearing and Neil 1999; USAID 1996; Budowski 1976). It is commonly referred to as a *symbiotic* relationship between conservation and development.

Ecotourism activist, academic and entrepreneur, Harold Goodwin, argues the case succinctly, promoting the potential for 'a symbiotic relationship' (Goodwin 2000). Ecotourism is, he argues, distinctive in that revenue and development arising from it are dependent on environmental preservation. As such, ecotourism is held to have the capacity to resolve the tension between development and environment generally associated with economic development.

Ecotourism is indeed a popular innovation in discussions of rural developing world development based broadly on the attractiveness of 'symbiosis' as set out by Goodwin and many others. It is sometimes suggested as a less damaging form of development by environmentalists who fear the developing world may be committing 'ecocide' through logging, or other activities that

Table 3.1 Introductory summaries for the five case studies

Organisation	*Principal aim*	*Relationship to ecotourism ICDPs*	*Structure and funding of organisation*	*Tourism within overall work*	*Comments*
WWF	Conservation	It seeks to influence the whole industry, but ecotourism ICDPs are one tool utilised in the context of the developing world to promote sustainable development.	Largest conservation organisation in the world, operates through country, region and project offices/many supporters worldwide, the WWF logo is well known. Funded by supporters, through international and multilateral bodies concerned with conservation, and by businesses.	A small part overall, but has developed in importance since the 1990s.	WWF highlights the importance of communities alongside conservation, but has been criticised, especially in the past, for placing conservation above community interests.
CI	Conservation	It seeks to influence the whole industry, but ecotourism ICDPs are one tool utilised in the context of the developing world to promote sustainable development.	International conservation organisation. Funded through international and multilateral bodies concerned with conservation, and by business and other donations.	A small part overall, but has developed in importance since the 1990s.	Founded in 1987 on basis of challenging ‘fortress conservation’ outlook. Exemplary of the new community-based focus for conservation. CI has a high profile as a pioneer of this approach.

SNV	To promote development in the developing world, principally operating in rural areas.	Seen as a tool for sustainable development in rural developing world.	Dutch-based, international development agency, work with multiple stakeholders and funders, funded mainly from national government donors.	Only one tool in its armoury, but has grown in importance.	Rural focus of its development work has meant that ecotourism ICDPs have emerged as an important strategy – it is regarded as a pioneer.
Tourism Concern	To campaign to promote greater social justice in the tourism industry/ emphasis on 'people' and 'community tourism'.	Strongly supported and promoted in the developing world as sustainable tourism development/criticisms of some conservation-oriented practice.	Small but influential campaigning membership organisation which supports financially/ also some project funding from governmental bodies and other NGOs.	Sole focus	Considerable success in influencing the general debate/has become a point of reference regarding ecotourism especially with its advocacy of 'community tourism'. It is exemplary of other campaigns, and expresses a widely held critical view of modernity in the form of mass tourism.
IYE	To synthesise experience and promote and integrate conservation and development via ecotourism.	Sole focus	Keynote conference, organised through UNEP and WTO featuring a range of NGOs of a conservation and well-being orientation.	Sole focus of the conference	The IYE represents a coming of age of ecotourism ICDPs, codifying their status as exemplary rural sustainable development through a UN-sponsored conference and its associated publications.

use up natural resources, in their struggle to survive (Cater 1994: 84). The UN IYE advocates ecotourism in Africa as a way of making 'the conservation of natural resources [. . .] mainstream to socio-economic development' (UNEP/WTO 2002a:11). Both arguments identify a 'win-win' situation between conservation and development. The community can earn money from tourists appreciative of the natural environment, and this money can support the community in their existing way of life, offsetting the desire for other forms of development that may be deemed less 'sustainable'. The direct benefits to the local populations concerned may include the opportunity to work in conservation, salaries paid from aid funds, revenue from ecotourism, and sometimes infrastructural benefits such as schools and medical facilities.

Yet, though material benefits from ecotourism are evident, they can only ever be, by their nature, very limited. While the literature on ecotourism emphasises the non-consumptive utilisation of natural resources (see, for example, Fennell 2003; UNEP/WTO 2002a; Scheyvens 2002), the traditional conception of development involves *transforming* the natural world for productive ends. Ecotourism's symbiosis between conservation and development is, it will be argued, a static one – it eschews transformative, thoroughgoing development, typically as 'unsustainable'. It fails to challenge, and in fact celebrates as 'sustainable living', the direct relationship that rural developing world communities have to their natural environment.

Symbiosis ties development to localised natural limits. But how does this come to be presented as sustainable development, as a positive innovation in rural development? What are the assumptions and premises behind this view? In Chapters 4 and 5 we look at the emphasis on community participation and tradition respectively, each of which present ecotourism as empowering, reflecting the desires and agency of the community. Following this, Chapter 6 examines the assumption of environmental fragility, an assumption that underpins the notion of localised natural limits and, it will be argued, ties the fate of local communities to these limits.

4 Community participation in the advocacy of ecotourism

Introduction

The term 'community participation' has a progressive feel to it – who could possibly object to greater participation? It suggests a greater level of control by and democracy for people – surely a laudable goal at all times and in all things. Certainly, the community participation agenda is a broad one in contemporary society. In the UK, for example, there is great concern with improving participation in elections and in the voluntary sector, and in the US the development of social capital through community involvement has become a prominent theme in political discourse.

In the developing world, too, 'getting local people involved' in projects for development and for conservation is a commonplace theme. According to one account, 'since the 1970s in many ways, community participation has become an umbrella term for a supposedly new genre of development intervention . . . [T]o propose a development strategy that is not participatory is now almost reactionary' (Tosun 2000: 165). It is perhaps precisely because of this latter sentiment that, as a concept, community participation is rarely subject to critical analysis. Instead, substantial critical studies tend to focus on the problem of *operationalising* the concept, rather than on the concept itself (e.g. Tosun 2000; Reed 1997). Precisely for this reason it is worth looking more critically at the often grand claims made for community participation in ecotourism by the NGOs featured in the case studies.

The implication of the call for greater community participation is often that it is more *democratic*, as it involves communities in decisions that affect their lives. It suggests a greater degree of *control* for the community over their destiny, rather than control being exercised from outside. Often this sentiment is articulated explicitly, too, through terms such as 'empowerment'. In this sense, the call for community participation in development is very much in the neopopulist tradition – it emphasises the role of communities in *their own* development.

This chapter establishes the centrality of community participation in the advocacy of ecotourism as sustainable development in the developing world, examines the case studies to ascertain their approach, and, with reference to the broader literature, provides an analysis of the claims made in the case studies. It will be argued that community participation is presented as a principle, and is associated with a progressive, democratic impulse, both in the literature on ecotourism in general, and in the case study material specifically. However, it is further argued that for the NGOs, community participation may be driven by other goals, principally that of conservation, and it can be in an important sense *instrumental* to these other goals.

Also, it is argued that communities are invited to participate only in the implementation of ecotourism projects, rather than in shaping the development agenda behind them, and hence real choices may be narrowly defined. The community's participation may be more of a pragmatic choice than an extension of democracy. The idea of community participation as *instrumental,* and its acceptance by communities as *pragmatic,* will be counterposed to the grander claims made both in the case studies and in the wider literature.

It is noted that the 'community' in 'community participation' is always envisaged as a *local* community, and often local initiatives are explicitly or implicitly regarded as progressive vis-à-vis participation and development at the *national* level. Indeed, criticism of grand development schemas at a national level, and the privileging of local development as a progressive alternative, are both key aspects of the neopopulist outlook that informs ecotourism ICDPs. However, it will be argued that local community participation, in privileging the local level, can be criticised for a failure to address adequately development at the national level.

Finally, the chapter questions the alternative and radical credentials of the community participation agenda, and whether it has the capacity to confer control over development in any meaningful sense.

The centrality of participation

As discussed in Chapter 2, the convergence of strands of thinking from development and from conservation have placed community participation centre stage. Indeed, this is clear when we consider the rise of sustainable development in contemporary social thought. Influential expositions of sustainable development argue for community participation as being of great importance. While remaining elusive in practice, it encapsulates the aspiration not only to combine conservation and development, but to engage communities and societies in this project. For example, the influential *Caring for the Earth: a Strategy for Sustainability* (IUCN 1991) lists one of its nine principles for sustainable development as to 'enable communities to care for their own environments'. Notably, the UN Conference on Environment and Development – the event that proved to be a watershed in establishing sustainable development as a rhetorical orthodoxy – put great emphasis on community participation (UN 1993), and was itself a striking example of the perceived need to involve communities and various stakeholders (many NGOs, large and small, were invited to the summit, although it has been argued that their actual participation was quite limited [Adams 2001; Mowforth and Munt 1998]).

More specifically, community participation is fundamental to neopopulist views on development, views which are very influential in the advocacy of sustainable development more broadly (Potter *et al.* 1999: 177–81), and that characterise the advocacy of ecotourism by the NGOs featured in the case studies. A typical neopopulist definition of community participation is that it should be about 'empowering people to mobilise their own capacities, to be social actors, rather than passive subjects, [to] manage the resources, make the decisions, and control the activities that affect their lives' (Cernea 1985, cited in Barnett 1995: 3). This definition emphasises control by the community – it is clearly *their* agency that is at the forefront of this formulation of development, not that of foreign governmental, commercial or non-governmental agencies. This sentiment is widely expressed in the literature on participation (Warburton 1998; Singh and Titi 1995; Stiefel 1994). It is a central feature of the outlook of many NGOs and was also prominently expressed at the watershed Rio Summit (UN 1993).

Participatory techniques in this neopopulist mould have evolved and become influential too. Participatory Rural Assessment (PRA), developed most notably by Chambers, emerged as a marginal idea in the 1980s, but grew in influence to become mainstream in rural development (Cornwall and Pratt 2003; Chambers 1997). Similar techniques are a concern of the NGOs involved in ecotourism ICDPs.

Neopopulist writers such as Friedman (1992) argue that development in the context of developing world states should embrace self-sufficiency, self-determination and empowerment, as well as improving people's living standards. The term empowerment in particular is ubiquitous in the discourse and, for France, applies to, 'individuals, households, local groups, communities, regions and nations' enabling them to 'shape their own lives and the kind of society in which they live' (France 1997: 149). However, in reality empowerment is almost always applied to local communities or individuals – certainly, ecotourism by definition involves small, localised projects and in such projects the participation of local communities is invoked as empowering, and as promoting control over development.

Community participation is absolutely central to the advocacy of ecotourism – all the case studies, as is evident in the summaries in this chapter, view it as a point of principle and as an intrinsic aspect of the projects themselves. It is universally agreed that such community participation is desirable, the outstanding issues being the *extent* of participation and the *form* it takes. This is especially so since the 1980s, a decade in which seminal publications such as Murphy's *Tourism: a Community Approach* (1985) and Krippendorf's *The Holidaymakers: Understanding the Impact of Leisure and Travel* (1987) established community participation as orthodoxy in the literature on tourism and development. All prominent authors concur. For Prentice (1993: 218), 'community involvement in tourism development has become an ideology of tourism planning'. In similar vein, with reference to the developing world, Mowforth and Munt argue that '[t]he debate is currently not one of whether local communities should be involved in the development of tourism to their areas, but how they should be involved and whether "involvement" means "control"' (Mowforth and Munt 1998: 103–4). This emphasis on participation mirrors its rise in development and conservation thinking more generally (Adams 2001; Cook and Kothari 2001).

Notably, community participation is considered as vital for achieving *sustainable* development (Scheyvens 2002). Indeed, in a thorough review of literature on community participation in tourism, Tosun even argues that 'a community approach to tourism development is a *prerequisite* to sustainability' (Tosun 2000: 617; my italics). The view that community participation is so important for sustainable development is based on the logic that it is the communities living in and around conservation areas who are best placed to manage the environment in a sustainable fashion. In relation to this Mowforth and Munt (1998: Ch. 6) usefully point out that sustainable development has become a 'socio-environmental category', embodying this relationship between people and environment, rather than being an environmental category per se.

It is also notable that the role of local community participation in establishing sustainable development is sometimes counterposed to the experience of mass tourism. Mass tourism, as an exemplar of modern, mass society, is often considered to have been too grand and impersonal to reflect the diverse cultures and views especially of villages in rural areas. For example, according to Brohman, 'developing countries may avoid many of the problems that have plagued past tourism [. . .] by involving diverse social groups from the popular sectors of local communities in decision making' (Brohman 1996b: 568). Here, Brohman presents local and small-scale initiatives as a partial antidote to national development schemas on a grander scale. Indeed, ecotourism has acquired a certain moral authority vis-à-vis mass tourism in debates on sustainable development (Butcher 2003a).

While there may in practice be a gulf between the ideas expressed in the literature and the reality of tourism planning, those directly involved in planning have bought heavily into the ethos of community participation too. For example, WTO tourism planner Inskeep has advocated community participation as essential to tourism planning (1991: 29), and elsewhere industry practitioner and academic Brent-Ritchie correctly predicted that resident-responsive tourism would become 'the watchword of tomorrow' (1993). International agencies as diverse as the WTO, the World Travel and Tourism Council (WTTC), World Bank, the UN, national development agencies and NGOs, have all adopted community participation as their own in general, or with regard to tourism in the developing world in particular.

Community participation, then, is widely supported and advocated with regard to tourism in the developing world (see also Fennell 2003; Scheyvens 1999; Hawkins and Khan 1998; Theopile 1995). Community participation in ecotourism can, it is held, increase the extent to which local communities have 'control'. It can 'empower' them, make them more 'self-sufficient', or give them 'ownership' over a project. These terms, and the neopopulist sentiments that lie behind them, are commonplace in the advocacy of ecotourism in the literature, and this is also evident in the case studies. Moreover, community participation carries an association with sustainable development and hence also a legitimacy and authority in development discourse.

A further important aspect of community participation is the assumption that the community should be a *local* community, as opposed to a *national* community. This is the corollary of the neopopulist emphasis on development conceived of at a local level, rather than a *national* level, at least in the first instance. Central to this view is what Sachs has called 'participatory planning and grass roots activation' (1979: 113). For Glaeser and Vyasulu (1984: 26), participatory development should mean that 'people who are affected by changes which they have decided are desirable cooperate voluntarily in the process of implementing the changes by giving them direction and momentum'. The authors here are referring to 'ecodevelopment', small-scale developments that encourage sustainable development on a local, rural and small-scale basis (ibid.). Formulations such as these posit participation as a local affair and implicitly prioritise local views over regional and national ones in development.

Yet dilemmas over differing local and national priorities are often a feature of development, both in the developed and developing worlds. For example, in recent years big dam projects in India, Turkey and China have involved the displacement of rural communities, but at the same time have great potential for the generation of electricity, and consequently higher living standards nationally. In the UK, infrastructural projects such as airport runways, reservoirs and motorways are often subject to substantial opposition locally, yet are deemed by many to be an important contribution to development and well-being at a national level. The neopopulist view consistently privileges the local over the national in such questions.

This privileging of the local is mirrored and magnified in the specific literature about ecotourism. Brohman puts this case clearly:

> Community based tourism development would seek to strengthen institutions designed to enhance local participation and promote the economic, social and cultural well-being of the popular majority. It would also seek to strike a balanced and harmonious approach to development that would stress considerations such as the compatibility of various forms of tourism with other components of the local economy; the quality of development, both culturally and environmentally; and the divergent needs, interests, and potentials of the community and its inhabitants.
>
> (1996: 60)

Here, the *local* community, not the nation, is clearly cited as the appropriate level to address a development that is environmentally and culturally benign.

It is *local* participation that is to be enhanced, and the *local* economy, rather than the national economy, with which tourism is to be compatible in this formulation.

The emphasis on local development is clear in Scheyvens' book *Tourism for Development: Empowering Communities* (2002). Scheyvens makes explicit that the locality is the most appropriate unit for development in terms of human well-being. She says of her book, '[i]t is not a book about how governments can extract the greatest economic benefits from encouraging foreign investment in tourism. [. . .] Rather, the interests of *local communities* in tourism development are placed at the forefront' (ibid.: 8; my italics). In her estimation, it is governments that benefit from a more traditional approach to development, whereas ecotourism ICDPs can be oriented towards local people in their communities, and are hence deemed to be what prominent neopopulist advocate Chambers refers to as 'good change' (Chambers 1983). This is typical of the neopopulist outlook on development – it presents large-scale development as beneficial to distant governments, with local community level development as holding out greater potential for *people*.

Other authors argue a similar point – that 'good change' should be organised around the local community's relationship with the natural environment. This is a popular, and populist, approach, and informs a great many studies on the impact of tourism on communities in journals such as the *Journal of Sustainable Tourism* and the *Journal of Ecotourism*.

So there is a strong sense in the literature on ecotourism and development, mirroring the more general neopopulist literature, that local community level development is the most appropriate spatial unit from which to address development, and that this might yield 'good' development, or a more 'sustainable' development. This is presented as progressive compared with the grand schemas of states, schemas typically proposing modernisation and transformation beyond the local, at the national level.

The following section gives an account of the views on community participation evident in the case studies, in the light of the general advocacy of community participation in the literature summarised above, and identifies some important themes from this. The relevant literature on this issue is quite diverse, and hence subheadings have been used within the case studies to help clarify these themes. The themes will be examined later on in the chapter.

The case studies and community participation

WWF

The centrality of community participation

In the case of the WWF, imparting a high degree of control to the local community is a central feature of its advocacy of ecotourism. Its emphasis on community is summed up in a position statement thus:

> Local communities reserve the right to maintain and control their cultural heritage and to manage the positive and negative impacts that tourism brings. Tourism should therefore respect the rights and wishes of local people and provide opportunities for the community to participate actively in decision making and consultations on tourism planning and management issues. Local traditions should be taken into account in buildings, and architectural development should be in harmony with the environment and the landscape. The knowledge and experience of local communities in sustainable resource management can make a major contribution to responsible tourism. Tourism should therefore respect and value local knowledge and experience, maximise benefits to communities, and recruit, train, and employ local people at all levels.
>
> (WWF-International 2001a: 3)

Here, WWF goes beyond the general rhetoric of community participation and specifies particular goals within it. These goals include the valuing of local knowledge and experience, development 'in harmony with the environment', the community's ability to 'control' their cultural heritage and the provision of economic opportunities for local people 'at all levels' (ibid.). Some of these will be examined in subsequent chapters, but it is notable that they are referred to here as very much part of the community participation agenda as a whole.

Community participation is also linked to 'sustainable resource management' (ibid.), and to 'responsible tourism' development (ibid.). As such it is cited as the appropriate level for development, or to be precise for the combined goals of conservation and development, to take place.

WWF, community participation, and GATS

WWF has sought to defend community participation in the face of the General Agreement on Trade in Services (GATS) legislation, which comprises the World Trade Organisation rules on international trade in services. These rules are widely interpreted as seeking liberalisation of the international trade in services, including the reduction of what may be viewed as barriers to entry, such as tariffs, quotas, investment controls and also participatory planning. GATS has been in force since 1995, and makes explicit reference to tourism advocating the '[r]emoval of measures imposed on tourism service providers regarding limitations on establishment, ownership and employment of personnel and remittances' (cited in Russell 2000: 91). WWF fears that such global liberalisation may end up restricting the ability of NGOs and host governments to work together in developing localities on the basis of what is distinctive about them – their culture and their environment – as GATS restricts the ability of governments to put conditions on investment (WWF-International 2001b: 3). Specifically, it argues that it may become more difficult for host governments, in conjunction with NGOs such as WWF, to advocate community participation as a condition of development (ibid.).

According to WWF's statement on this issue:

> [i]t may become harder for governments to put in place regulations to guarantee local participation in tourism development or to preserve their natural resources (e.g. a hotel chain may challenge a wetland classification for a resort by questioning the necessity for such a restriction). There is a risk, therefore, of a chilling effect of GATS on new regulations aimed at promoting sustainable tourism.
>
> (WWF-International 2001b: 3)

On this basis WWF has argued strongly for GATS to take account of measures necessary to develop what it considers to be sustainable tourism in the developing world – tourism based in communities around their existing environmental and cultural assets, in which community participation is prominent.

WWF as facilitators

WWF tourism expert Justin Woolford emphasises the importance of community participation to WWF thus:

> It is quite difficult to separate ecotourism and community based tourism as far as we are concerned and as far as our projects are concerned. We'd subscribe to The International Ecotourism Society view about travel to natural places that is beneficial to people and to the environment [. . .] we'd subscribe to that but maybe go a little bit further and say *we'd like it to be determined by local people*, rather than it just being beneficial for them.
>
> (Woolford 2002; my italics)

That the community, not outside agencies or the WWF itself, 'determine[d]' outcomes is a consistent feature of the advocacy of community participation. Implicit in the above quotation is the role of WWF as a *facilitator* of development, a process that it rationalises as enabling developing world peoples to 'preserve *their* natural resources' (ibid.; my italics). The rationale here is that the natural environment and culture are *of the community*, and the role of WWF is to facilitate the community in benefiting from these resources, while also conserving them.

The facilitation role of WWF, and the controlling interest of the community, are prominent in WWF's conceptualisation of community participation elsewhere too. Gaynor Whyles, WWF's European Policy Officer, referring to projects on Albania's Mediterranean coast, asserts that '[t]he Albanians are in the driving seat of [the] project. Through a process of participatory development – probably the most important part of this project, *they* will use the skills of our project team to realise *their expectations*' (cited in Farrow 1995: 9; my italics).

Further, Cherry Farrow, Communications Officer for WWF, argues that:

> WWF has long experience in developing integrated conservation and development projects that aim *to assist rural communities to manage their natural resources wisely*, promote rural development and solve environmental problems through the use of both formal and community education.
>
> (Farrow 1995: 9–10; my italics)

Clearly, WWF's role here is presented as primarily that of facilitation of the community.

Another prominent example of this rationale is the Communal Areas Management Programme for Indigenous Resources (Campfire) programme in Zimbabwe, a longstanding and often discussed ecotourism ICDP in which WWF is centrally involved (Campfire undated). WWF argues that:

> [f]rom the very beginning, *the indigenous people of these communal areas became the driving force behind the Campfire Programme.* The Tonga, Venda, Ndau, Ndebele and Shangaan people of Zimbabwe became responsible for managing their own natural resources and were able to retain significant benefits at the village and ward levels.
>
> (WWF-UK undated d; my italics)

An instrumental rationale for community participation?

However, despite claims such as these to be simply facilitating the empowerment of the community, there is a strong sense in WWF's literature that this empowerment provides an important point of contact with the community through which distinctly conservation-oriented values can be promoted. One document, referring to projects concerned with carnivore conservation, alludes to this as follows:

> Tourism can increase a local community's sense of pride in, and value placed on, large carnivores as a visitor attraction. A community's sense of ownership and involvement can open the way for discussion about conservation measures or development plans, which may directly improve their livelihoods.
>
> (WWF-UK 2000: 6)

What might be regarded as the *instrumental* nature of community participation – that participation is a means to an end rather than the expression of community wants per se – is also evident in WWF's *Guidelines for Community Based Ecotourism Development* (WWF-International/Denman

2001). WWF argues that ecotourism can: 'provide a more sustainable form of livelihood for local communities; encourage communities themselves to be more directly involved in conservation; generate more goodwill towards, and local benefit from, conservation measures such as protected areas' (ibid.: 4).

A key question posed by the author elsewhere in this report is: 'What type and level of incentive might be needed to change attitudes and actions in order to achieve worthwhile conservation benefits?' (ibid.). So here community participation is a way of engaging the community in a dialogue as to how they might adopt ecotourism as an environmentally benign form of development. Specifically, the participation process, to be successful, has to build up 'goodwill' towards conservation measures, and to 'directly involve' the community in conservation (ibid.). Community participation thus could be argued to be instrumental to the wider and prior conservation goals of WWF.

The limits to community participation – WWF and tourism in the Arctic Circle

WWF's work in conservation in the Arctic provides a stark example of the limits to community participation when the community chooses something other than integrated conservation and development. WWF states within its *Ten Principles for Arctic Tourism* (WWF-International undated a: 3) that: 'Local involvement in the planning of tourism helps to ensure that tourism addresses environmental and cultural concerns. This should maximise benefits and minimise damage to communities. It should also enhance the quality of the tourism experience' (ibid.).

Elsewhere, it argues strongly for a common cause for environmentalists and the tourism industry against oil developers in the region. WWF advises tour operators to:

> [p]romote maintenance of large, undeveloped areas of the Arctic. The undeveloped regions of the Arctic have a unique value, and are one of the primary reasons why tourists come to the Arctic. This will be undermined by roads, pipelines and other kinds of unsightly large-scale development that fragments the environment.
>
> (WWF-International undated a: 5)

All this assumes that the local communities share the same 'environmental and cultural concerns' (ibid.: 3) as WWF. But in the case of many economically poor communities in the Arctic, the prospect of oil revenues has created tension between local wishes and the outlook of the conservationists at WWF. For example, while a proposal put forward by the US government to allow drilling in a small part of the Arctic National Wildlife Refuge (ANWR) drew principled opposition from WWF and other environmentalists, many reports on the issue have found that there is widespread support for the plans from the various indigenous Innuit peoples living in the region (e.g. Lister 2001a and 2001b).

The problem is, in this case, that WWF's aim to preserve the wilderness and biodiversity is not shared by the majority of the local and indigenous people, who see oil revenues as offering considerable benefits (ibid.). Moreover, tourism, as an alternative form of development deemed 'sustainable' by WWF, cannot claim to offer the same development prospects as oil. In this case, for WWF, the conservation imperative has taken precedence over the wishes of the community, and WWF has continued to campaign internationally against further oil development and proposes tourism as an environmentally benign alternative, both in the Arctic and elsewhere, in spite of the views of the community (WWF-International 2002).

Communities can choose, apparently – WWF's *Ten Principles for Tourism in the Arctic* (WWF-International undated a: 3) includes the statement that developments should 'respect the rights and wishes of local and indigenous people' – but when those choices conflict with conservation, then they may be deemed illegitimate. A conservation imperative thus sets the terms of, and limits on, community participation.

The above example suggests that community participation is ultimately conditional on an acceptance of goals that are established prior to the involvement of the community. Further, this summary of WWF's position on community participation also suggests that it has a significant *instrumental* side to its character – it is, in part at least, an instrument to establish allies for conservation, and to educate and incentivise the local population accordingly.

Local community participation and the national dimension of development: the example of Campfire

It is also clear from the summary thus far that community participation is invariably envisioned as at a local level, typically the level of the village or villages. This corresponds with the spatial level at which conservation is organised – typically rural, undeveloped areas, relatively sparsely populated, in which people are often engaged in subsistence agriculture, organised through village structures. But it is worth considering the relationship between this and more traditional national conceptions of development and, indeed, democracy. The Campfire project in Zimbabwe provides an exemplary and revealing case.

WWF was one of the originators of Campfire as far back as 1984, and has played a central role in its development (WWF-International undated c: 1). Campfire focuses on communal land areas which have retained significant wildlife resources (constituting 14 per cent of Zimbabwe's landmass), rather than state owned and managed game reserves (ibid.). Often these are lands on the periphery of areas rich in wildlife, including species protected under CITES legislation, and are the home to some of the poorest communities in Zimbabwe. The project effectively encourages wildlife conservation, but on the basis of establishing opportunities for the local communities to benefit from it. The two stated aims of Campfire are: to conserve natural resources in the communal areas; and to increase income-earning opportunities in poor communities (Campfire undated). Income is derived from ecotourism and controlled hunting tourism, alongside other related activities such as photography and fishing. Through this, the communities have been able to derive economic benefits (for example, income from working in tourism or on conservation-related activities, or from small capital projects such as schools and health centres) from supporting conservation (either through refraining from hunting or from working on conservation related activities) (Logan and Moseley 2002; Child 1996). Campfire is a clear example of the incentivisation of conservation among this impoverished rural community.

Campfire is funded through a variety of governmental and non-governmental aid agencies, including WWF. It involves a complex system of administration, the shape of which is itself revealing. The Zimbabwean Department of National Parks and Wildlife Management (DNPWM) works closely with a variety of NGOs including the WWF. At a national level the Ministry of Local Government, Rural and Urban Development (MLGRUD) is concerned with income generation and is legally responsible for the management of wildlife within the limits set by the DNPWM, whose primary role is conservation. At a local level, Rural District Councils (RDCs) play a key role in implementation (Murombedzi 1999; Plan Afric 1997).

The process is driven from the northern NGOs, such as WWF, their conception of wildlife and the communities' relationship to it – it is these NGOs that provide the finance for Campfire. While it is true that the Zimbabwean central government is centrally involved through the DNPWM, there is in fact a special Campfire unit within this department which has tended to bypass local DNPWM offices (Plan Afric 1997: 4). The NGOs play a major role in this unit. This is in keeping with a process described thus: 'The general trend in Campfire has been to set up its own set of structures, which operate in conjunction with, but separate from, the existing central and local government structures' (ibid.).

Although community participation is well established within the Campfire literature, some have questioned the reality (McIvor 1997). One report suggests that the aim is 'not to allow local communities to choose what to do with "their wildlife", but to teach them how to manage it in the manner DNPWM (the Zimbabwe Department of Parks and Wildlife Management) sees fit' (Plan Afric 1997: 4). Also, the funding authorities have the sanction of withdrawing Campfire status if their policies are not complied with. According to one source Campfire is less 'co-management', as claimed, and more a case of 'persuasion' (ibid.). Ultimately that power to persuade is strong, backed up by the financial authority of Campfire's principal partner, WWF. It is not fanciful to suggest that poor communities will accept its limited benefits on the terms available, rather than questioning this at the risk of cutting off these benefits.

Promoting local participation can involve modifying or changing the authority of the community over their land, which inevitably involves an engagement with national law. For WWF, community-based ecotourism requires 'where possible a strengthening of the legal rights and responsibilities of the community over land, resources and development' (WWF-International/Denman 2001: 10). This is held to be important regarding communities' influence on the shape of economic activity (ibid.). In the early days of Campfire, lobbying for greater indigenous land rights was an important factor in establishing the Campfire modus operandi.

The Campfire scheme is a good example of the benefits that can accrue from attracting ecotourists. However, it also reveals that orienting development aid at the level of the community opens up questions relating to governance, power and the nature of development. These questions remain unaddressed by WWF in its literature.

Conservation International

CI also puts great store in community participation. Tourism expert Jamie Sweeting sees his original role with the organisation as to 'help communities develop ecotourism' (Sweeting 2002). As such, his role is presented as being a facilitator of development, the clear implication being that it is the communities themselves who play a leading role in endogenous development, a key feature of the neopopulist outlook. There is a strong emphasis in its literature on community participation, and this is reflected in its practice too.

However, as with WWF, the conservation aims of CI strongly influence the terms of community participation. CI's biggest single funder is the United States Agency for International Development (USAID) (CI 2003), the aid arm of the US government. USAID supports projects that claim to integrate conservation and development activities in many developing world countries through CI (USAID 1996). Such projects, USAID claims, provide alternatives to encroaching into protected areas to hunt, log and farm (ibid.). Furthermore, through community participation, 'a new group of stakeholders with a vested interest in protecting parks' is created (ibid.: 1). It is clearly important for them to offer benefits to host communities, as 'potential local resistance to setting aside forest and fishing areas for conservation can often be softened by employment and income producing opportunities ecotourism can generate' (ibid.). This suggests that sponsorship of ecotourism is after all to do with environmental imperatives, and that the small economic benefits to communities are instrumental to this aim – to clear the way for its acceptance within developing world communities.

In similar fashion, CI itself argues that '[a]ll projects need to integrate the conservation of neighbouring ecosystems with the creation of economic opportunities for local residents . . .'(CI 1999: 1). Furthermore, 'the development of an ecotourism project depends on building a local constituency that has a vested economic interest in protecting their natural resources' (ibid.). Community participation enables the development of this constituency for conservation, and is shaped by this overarching aim.

So, as in the case of WWF, community participation amounts to participation in a pre-existing agenda, rather than in determining the agenda, and hence is instrumental in character. Another parallel with WWF is the importance of *local* development, in which the locality corresponds to the interface between human populations (often rural villages) and the environment within 'biodiversity hotspots'. Development issues elsewhere, and beyond this spatial level, are not within the remit of the conservation organisations. Hence, local community participation is a product of the geography of these specific interfaces, with little reference to a wider development agenda.

SNV

The centrality of community participation

For SNV, the participation of local people is 'an important principle in developing a sense of ownership of the project' (SNV/Caalders and Cottrell 2001: 32). It claims to place a strong emphasis on social mobilisation and local governance linked to community development (ibid.). Indeed, participation by the local community is viewed as a prerequisite for projects if they are to meet the aims of SNV. Participation 'should be incorporated into the very beginning of a programme to maintain responsibility, interest and ownership of the development activities within the local actors' (ibid.: 6). However, it is accepted by SNV that 'community management should not become a dogma' (ibid.: 33), as clearly private sector expertise and the ability to tap financial resources from elsewhere are important considerations too.

SNV's emphasis is clearly on the neopopulist aim of generating development that is endogenous, from within the community, rather than exogenous. For example, in Botswana, SNV regards 'community based tourism not as an end in itself but as a means towards empowering poor communities to take control over their land and resources, to tap their potential, and to acquire the skills necessary for their own development' (SNV/Rozenmeijer 2001: 7). This document, discussing SNV's experience in Botswana, goes so far as to assert that, through ecotourism ICDPs, people 'have the potential to control their own development process' (ibid.: 55). Here, the emphasis is put on the community – *their* potential, *their* development. SNV's role is presented as being to empower communities towards this end.

Further in this vein, there is a strong focus on 'self help' in the SNV literature on tourism (SNV/Caalders and Cottrell 2001: 20). Its aim is to 'help people better understand the opportunities *they* have to improve *their own* situation' (ibid.; my italics). As such, SNV presents itself as a facilitator of the agency of the community, agenda-less, rather than a protagonist in the ideological issue of what constitutes 'good' development.

The role of facilitator is elaborated upon by SNV thus: 'To facilitate the development of a truly representative community organisation that can deal with the different interests in resources use, and can ensure an equitable distribution of benefits, requires the involvement of an outside agency (NGO, government department, consultant)' (SNV/Rozenmeijer 2001: 57). This is based on a clear understanding that communities can contain divisions based on class, gender, age and power (ibid.). Further, the NGO 'plays the role of a broker at different levels and links the community with the other stakeholders' (ibid.: 61). Notably, it adds that '[a]s long as "community structures" have to manage "community resources" some form of light touch facilitation may be required on a permanent basis' (ibid.: 61).

SNV has written at some length on the importance of community participation, and the concept of 'community' itself (e.g. SNV/Caalders and Cottrell 2001). It argues that a community tourism project can only succeed when the 'community has been clearly defined by all residents and a truly representative organisation has been built that is accepted by all stakeholders in the area' (SNV/Rozenmeijer 2001: 32). Indeed, SNV goes to great lengths to involve the community in its projects. It advocates a thorough process to ensure that community participation is not piecemeal, and that it is as true a reflection of the community's wishes as possible (SNV/Rozenmeijer 2001: 23). Specifically, it has adopted a particular planning methodology, 'Appreciative Participatory Planning and Action' (APPA), at community level, and claims that it gives the community 'the possibility to plan their own tourism related village development plans' (SNV/Caalders and Cottrell 2001: 32). The method is 'geared towards practical solutions' with regard to the aims of the project (ibid.).

Ownership, aid and natural capital

An important theme is that developing world communities should, and should be able to, take responsibility for their own development, and that community participation is a means to this end. For example, prominent within its literature on ecotourism ICDPs is the notion of 'ownership', which is based on 'a recognition that national governments and local actors should take overall responsibility for their own development, which leads to empowerment' (SNV/Caalders and Cottrell 2001: 32). In similar vein, in a general statement in its annual report of 2001, SNV argues that 'ownership is the underlying principle. Everyone has the fundamental right to determine the course of his or her own life, his/her own country. This is reflected in our working alliances, where deference is our watchword' (SNV 2001: introduction).

SNV argues that community participation can also play the role of strengthening the community itself, who may be cohered around the management of local resources. It adds that:

> [t]here is more to CBT (Community Based Tourism) [. . .] than money and jobs. Positive changes take place that can be noticed, but these are

> difficult to measure in quantitative terms. These changes are important, because they are the foundation on which communities can manage their natural resources in a sustainable way.
>
> (SNV/Rozenmeijer 2001: 54)

In a sense, then, community participation is not just narrowly a means to an end, but is part of a process of mobilising the community around integrated conservation and development.

The view that local communities should take responsibility for projects is justified through criticism of the propensity for aid money to be viewed by recipients as 'handouts' from third parties (SNV/Rozenmeijer 2001) – SNV is critical of NGOs which promise funds, but neglect the primacy of the agency of the community. It argues that, in the case of Tanzania, '[t]oo much money falling from the air has created a social environment of depending on outsiders. The overwhelming donor presence in D'kar has negatively affected ownership and commitment to the project, and the motivation to work and learn' (ibid.: 45). SNV argues that it is vital for the community to take control and not to expect '[t]oo much money falling from the air' (ibid.). Implicit here is that too much aid may compromise 'ownership', as it will involve outside agencies – commercial, government or NGO – in proposing change to the community's way of life. In place of this, SNV places an emphasis on the non-consumption of natural capital in development – the ability of communities to derive development on the basis of conserving their environment and their culture, rather than through a process involving greater investment and change (ibid.).

This criticism of 'too much' aid is remarkable when one considers the dearth of development aid to the developing world generally. Landmark events concerned with development, such as the *Brandt Report* of 1980 (Brandt 1980) and the United Nations Conference on Environment and Development of 1992 (UN 1993), have laid down minimum commitments for aid to the developing world, commitments that have rarely been met. Perhaps the problem is not 'too much money falling from the air', but rather the dearth of development aid itself. Equally, the lack of 'commitment' and the culture of 'dependency' cited by SNV may have more to do with a distance between the priorities of the NGOs and the aspirations and political and economic realities in the recipient communities.

Differences and similarities between SNV and the conservation NGOs

SNV, as a development agency, is not beholden to a general conservation aim, as could clearly be argued in the previous two case studies. However, the anticipated outcome of community participation is similar. SNV argues that 'local level participation is essential for achieving the global goal of sustainable development' and, indeed, 'sustainable environmental management can only occur where active local level support exists' (SNV/Caalders and Cottrell 2001: 12). Sustainable development in rural areas is consistently interpreted as development principally on the basis of natural capital by SNV (SNV/Caalders and Cottrell 2001; SNV/Rozenmeijer 2001; SNV/de Jong 1999). Indeed, SNV explicitly favours this on the basis that it does not involve the need for new capital investment that may be deemed unsustainable for the environment, and may not be amenable to community participation to the same extent (SNV/Rozenmeijer 2001). Given these limits, it is hard to envisage rural development on any other basis than that of conservation, and trying to make conservation pay – which is precisely the basis of ecotourism.

This similarity in the expected outcomes of community participation between the conservation NGOs and SNV is strongly evident in SNV's literature. It argues that tourism can 'help justify and pay for the conservation of important natural areas and wildlife because these are attractions for tourists' (SNV/Caalders and Cottrell 2001: 43). Further, ecotourism can '[i]ncrease environmental awareness when residents receive jobs and income from nature tourism and observe tourists' interest in conservation, then they realise that protecting the environment is important' (ibid.).

So, although the instrumental nature of community participation is less clear in the case of SNV, the choices that would make a participation process democratic to a greater extent are no more in evidence than in the previous two case studies. Development based on existing, local, natural resources is deemed sustainable development in the rural developing world and, working within these limits, it is hard to envisage much beyond the small-scale ecodevelopment proposed by the conservation organisations. Ecotourism becomes a pragmatic option for communities.

Local participation and the national dimension of development

SNV does have a very clear perspective on the relationship between local participation and the higher levels of governance, at a regional and national level, and this is in some contrast to the conservation NGOs in the previous two case studies. SNV's discussion of participation is also linked to a wider recognition of the importance of developing governance – again something that is far less evident in the other case studies.

Illustrative of this is that SNV seeks to 'foster collaboration between actors at the meso level and [to link them] to higher national and international policy and institutional levels' (SNV 2001: 6). Meso level actors are seen as those organisations, both governmental and non-governmental, that link the local level to the national, and would include trade associations and trade unions. SNV considers these links between micro, meso and macro level organisations as important, and recognises that community participation must relate to other, pre-existing structures of participation and governance (SNV 2000: 13).

Further, it seeks to develop 'nodal points' linking the various actors and levels of government (SNV/Caalders and Cottrell 2001: 33). This emphasis on governance is part of a wider 'good governance' agenda that emerged in the 1990s, which is in part a response to the perceived failures of aid in the past to filter down to where it is most needed (Abrahamsen 2001). Within this agenda, decentralisation is a key component (ibid.).

However, while SNV has a clear and developed approach to governance, and the relationship between different levels of government, it also sees the promotion of the local community and community participation as 'a tool to readjust the balance of power and reassert local community views *against those of developers or local authority*' (SNV 2001: 12; my italics). Empowering the community therefore can involve a challenge to higher levels of governmental authority.

SNV's stance is strengthened by its argument that poverty is linked to access to power and resources at a community level. It puts an emphasis on meso level organisations to link communities with higher levels of governance in order to facilitate this passing down of power from national to local levels. For example, its multi-annual plan for 2001–3 argues for 'a genuine devolution of resources and authority' to 'create opportunities for local communities, traditional leaders, private sector operators and NGOs . . .' (SNV/Schuthof undated: 6). Further, it argues that the development of 'democratic local governance systems are a precondition for effective poverty reduction strategies' (ibid.).

The need to devolve power and resources can extend to ensuring a proportion of tourist expenditure on projects goes directly to the community. In the case of its Tanzanian Cultural Tourism Programme, SNV charges a development fee – a small fee added to the tour price that is explicitly to aid development. This fee is likened to a tax. 'What we are doing is comparable to taxation: we invest in achieving community objectives. The only difference is that this money is used in the right places, and sadly that cannot be said of all tax money in Tanzania' (SNV/de Jong 1999: 19).

A practical expression of SNV's promotion of local community participation is the case of the Bushmen in Botswana. The Bushmen are a marginalised group within Botswana. SNV has tried to address their poverty, and increase their claim-making power in relation to the government (SNV/Rozenmeijer 2001: 17). This has included lobbying for and obtaining indigenous land rights for the Bushmen over their habitat in the west of the country, to facilitate ecotourism (ibid.).

SNV's position clearly argues for a promotion of *local* development and participation relative to the national level. Yet the nation remains an important, if not the most important, spatial level at which to conceive of development. SNV's promotion of ecotourism as sustainable development in the rural developing world does not seem to consider the potential tensions between legitimate national priorities and localised ecotourism ICDPs.

Tourism Concern

Tourism Concern's critical stance, and its alternative, 'Community Tourism'

A lack of community participation is the central theme in Tourism Concern's criticisms of tourism's role in development. In the editorial of the Summer 1995 edition of its magazine *In Focus*, titled 'Local Participation – Dream or Reality' (Barnett 1995), it argues that, 'local communication has become a "buzz word" in the development field' (Barnett 1995: 3). The editorial is critical of the breadth of the rhetoric of community participation, alongside the reality of so few good examples. Barnett problematises the issue, arguing that 'tourism must recognise the rights of residents to be involved in its development and management. Without this [. . .] tourism cannot be equitable and have a long term sustainable future' (Barnett 1995: 3).

Community Tourism has emerged within Tourism Concern as the favoured option to bring about these high minded ideals, and the term has become quite influential within the wider literature on ecotourism. Community Tourism was formalised through Tourism Concern's influential publication *The Community Tourism Guide* (Tourism Concern/Mann 2000). In formalising and championing 'Community Tourism' as a form of tourism that puts 'people first', Tourism Concern has done much to challenge the claims of some NGOs with regards to the latter's actual practice in the area of participation. Tourism Concern's quarterly magazine, *In Focus*, features such criticisms regularly. However, its Community Tourism agenda also suggests a similar instrumental approach to participation. As *The Community Tourism Guide* states:

> If conservationists want [communities] to say 'no' to harmful development, they must offer them alternative means of feeding their families. Tourism may be that alternative. In many places, tourism is a central pillar of emerging alliances between local communities and conservation organisations.
>
> (Tourism Concern/Mann 2000: 27)

This is certainly the case – conservation organisations have developed such alliances, alliances that offer limited economic benefits on the basis of environmental conservation. Tourism Concern's approach is to argue trenchantly for the organisations to deliver in terms of community participation and the consequent distribution of economic benefits to the local communities. But it shares the view that thoroughgoing, transformative development is 'harmful development' (ibid.) in the rural developing world, and hence the scope of participation becomes, by default, small-scale, nature based development that is not 'harmful'.

Community tourism is considered by many to be the state of the art in ethical tourism. *The Community Tourism Guide* offers a brief analysis of this type of tourism and lists holidays that conform to it (Tourism Concern/Mann 2000). It defines community tourism simply as 'tourism that involves and benefits local communities' (ibid.). The guide goes on to argue that '[i]t is only by putting people at the centre of the picture that true conservation solutions will be found' (ibid.). This is revealing – conservation remains the aim, but local communities have to be 'at the centre' of projects to bring this about. *The Community Tourism Guide* clearly expresses the argument for integrating conservation and development through ecotourism.

But what if a community did not want to put the author's 'true conservation solutions' first? What if people prefer to leave the community in search of more lucrative jobs in hotels and in the cities? What if they view their culture as restrictive? What if they want to break away from the poverty in their community? Other prominent advocates of community participation in ecotourism ICDPs have noted that the community may not accept ecotourism's premises (e.g. Weaver 1998: 15), and this is also accepted by Sue Wheat, prominent and longstanding Tourism Concern member, and assistant editor of *In Focus* (Wheat 1994).

With its emphasis on 'community', Community Tourism certainly seems to provide answers for conservationists confronted with the accusation that they are only concerned with the environment. However, it offers a conception of development constrained by an imperative to integrate it on a localised basis with what are accepted as pressing conservation needs.

The congruence between community participation and smallness of scale

A theme in Tourism Concern's literature relevant to ecotourism ICDPs, and common to the other case studies, is an emphasis on small-scale development as preferable. One reason for this is that rural communities themselves are small in scale, and therefore to engage the community in their own development requires small-scale projects. Wheat articulates the rationale for small-scale development, but also hints at the limits of this approach, as follows:

> Most . . . [alternative tourism] . . . initiatives are small. Schumacher coined the inimitable phrase 'small is beautiful', but small can also be insignificant, and whilst many alternative tourism initiatives may be well meaning and fulfil the JPS criteria (Just, Participatory, Sustainable), they are unfortunately merely a 'drop in the ocean'. Even if the number of such projects increased dramatically it is not likely they could increase enough to cater for the scale of today's demand for holidays. And of course, if they did, they would no longer be small – and no longer beautiful – or JPS.
>
> (Wheat 1994: 2)

The problem, then, is that community participation, and sustainable development, according to this formulation, are only conceivable on a small scale. Wheat proceeds: 'Small operations, if successful, can end up big. And all too often any egalitarian, community minded principles get squeezed out, and winners and losers start emerging' (ibid.). The logic here is that community participation requires small-scale development. To go beyond this prejudices community participation, and its corollary, sustainable development. It is a logic that seems to implicitly limit both how development is conceived of, and, potentially, development itself.

A scale for gauging community participation

Tourism Concern, although a campaigning organisation and therefore not directly involved in the practicalities of implementing ecotourism ICDPs, has had much to say on how community participation can and should proceed. Tourism Concern's Mark Mann cites his own three-point scale for gauging the level of community participation in various tours. Whereas 'Responsible Tours' provide direct benefits to local communities, 'Community Tours', at the other end of the spectrum, are initiated and managed from within the community. Between these two lie 'Partnership Tours', where outside agencies are able to assist with skills and business knowledge (Tourism Concern/Mann 2000).

In the latter case, the partners are likely to be NGOs, and this is the case with many of the tours cited in *The Community Tourism Guide* – the guide includes projects funded by WWF, CI, SNV and a range of other development and conservation NGOs. Yet it is rare that such agencies simply offer business assistance. As tentatively argued in the earlier summaries, and developed later in this chapter, aid is tied to *particular* conceptualisations of development, and the community do not participate in establishing these. Assistance given is premised upon the NGOs' ability to shape the developments themselves.

Mann's 'Community Tours' (Tourism Concern/Mann 2000) appear to be truly neopopulist in the sense that they emerge from the community themselves. Yet, given the poverty of the rural developing world, where aid and assistance is normally a prerequisite to develop projects that link up with Western (tourism) markets, community tours of this kind are hard to envisage, a point accepted by Mann (ibid.: 31).

Concluding comments

Tourism Concern is distinctive as a campaigning NGO around issues raised by tourism development. As such, this rhetorical advocacy of community participation is relatively unrestrained by an agenda linked to a specific environmental imperative. However, there is clearly a general acceptance that development must be tied to conservation within the rural communities hosting projects, be it on terms as favourable as possible to the communities concerned. Further, the promotion of small-scale development, as a point of principle, linked to the imperative for sustainable development, reinforces a bias towards development based on the non-consumption of natural capital. Hence, perhaps Community Tourism is less of a critique of the 'conservation first' approach than it at first appears to be.

UN IYE

An emphasis on community participation

The UN *World Ecotourism Summit Final Report* (UNEP/WTO 2002a) is the product of input from numerous NGOs, both those from a conservation and a development bent. These include the other NGOs featured in this case study – CI, WWF, SNV and Tourism Concern. It reflects the centrality of community participation in the advocacy of ecotourism. It is also clear that this participation relates to the terms on which ICDPs are implemented – that they *should* be prioritised in rural development is not questioned.

Both the emphasis and limits on community participation are evident in the *Final Report.* As a document put together by a process involving many varied governmental and non-governmental organisations, including the other case study organisations in this study, it is a very significant contribution to the advocacy of community participation.

Community participation in ecotourism planning is insisted upon in the IYE documentation – it is considered central to ecotourism, if the latter is to be sensitive towards the local culture. For example, the *Quebec Declaration,* a summary of the main recommendations from the IYE, refers to 'the right to *self determination* and *cultural sovereignty* of indigenous and local communities' with community participation the means to this laudable end (UNEP/WTO 2002b: 66; my italics). On the face of it, this emphasis on factoring in to tourism development the community and their culture seems admirable and democratic.

The instrumental nature of community participation

Yet the purpose of the participatory process appears prescriptive. The IYE *Final Report* asserts that '[p]articipatory processes should be used to educate people about the value of biological and cultural diversity in ecotourism development, and on how they can both conserve and derive benefits from natural and cultural resources' (UNEP/WTO 2002a: 82). Hence, participation here is not about establishing the will of the people per se, but is instrumental in promoting a particular view of conservation and development.

This view emanates from the funders – NGOs and development agencies – and funding is invariably conditional on an acceptance of the terms of the projects offered. There is no facility for rural communities to decide on a purpose outside of this, to use the funding in an alternative way. Hence, in spite of the prominent neopopulist rhetoric of 'consultation' and 'participation' that characterises the IYE, ecotourism is substantially an imposed agenda here, and empowerment of the community is limited to how that agenda might be implemented. This is made explicit in the *Final Report*, which argues that ecotourism 'provid[es] a source of livelihood for local people which encourages and empowers them to *preserve the biodiversity of their local area*' (UNEP/WTO 2002a: 43; my italics). Of course, the assumption here is that the local people share this aim, and all that needs to occur is for them to be empowered to bring it to fruition.

Formally, there is choice, as through 'participative planning mechanisms', communities are able to 'opt out of tourism development' (UNEP/WTO 2002b: 67). Yet to do this would mean opting out of benefiting from the project funding or investment at all. For economically impoverished societies, this is hardly a meaningful choice. It is at best an extremely limited form of control over development, while the substance and trajectory of development is established elsewhere.

Local community participation and the national dimension

A striking section in the *Final Report* sets out a clear agenda on the trajectory of ecotourism ICDPs. The report includes a 'clear message from delegates' that 'donor agencies should provide more schemes which channel assistance directly to enterprises and communities rather than through national governments' (UNEP/WTO 2002a: 47), and that, as a 'principle', ecotourism should 'allow local and indigenous communities, in a transparent way, to define and regulate the use of their areas at the local level' (UNEP/WTO 2002a: 2). Local community, then, is the spatial unit at which ecotourism addresses development as well as conservation. Moreover, this empowerment of the community explicitly involves the disempowerment of the national government.

Apart from the above, the IYE, throughout its lengthy *Final Report*, does not address the relationship of this local development and local participation to national development and participation. How local initiatives contribute to national development, in a world in which nation states are the pre-eminent political actors, is an unasked question throughout the IYE documentation. The relationship between local community consultation and sovereign structures of government within states is rarely examined, and certainly unexamined in the IYE documentation. Instead, there is a consistent emphasis on the local. However, in explicitly advocating aid to rural areas on a localised basis, sovereign national structures of governance may be affected. Given that the aim of aid ultimately should be to make the need for aid redundant, and for countries to function and thrive without intervention, issues of governance and sovereignty remain surprisingly conspicuous by their absence. *Political* sustainability might be a category worth exploring in this respect.

Summary

A number of themes are clearly evident in the promotion of community participation by the case studies. First, the emphasis on participation, and the claims made for it, are both striking. In the literature examined there is a clear sense that community participation is a fundamental principle, and that it enables 'empowerment' of the community, or that it confers 'control' onto them. There is also often an association with sustainable development. These are quite grand claims, often made with little or no qualification, and, as such, deserve close scrutiny.

Second, community participation, elevated as it is to a fundamental principle, suggests that the NGOs' role is one of *facilitation* of the community in realising the latter's wishes. Indeed, four of the case study summaries directly refer to the role of NGOs as facilitators of the community, while all of them stress this role indirectly. Facilitation implies a disinterest on the part of the facilitator as to the outcome of the participation process. 'Empowerment', too, suggests that it is the community which is gaining a substantial degree of power or control, and that the NGOs' role is to bring this about. In this fashion, the NGOs featured in the case studies present themselves as catalysts for community democracy, rather than exerting a powerful external influence on the direction of development.

Third, although their role as facilitators of the community is a common theme, it is evident that there is clearly a prior agenda attached to ecotourism projects, rather than a disinterest as to where participation may lead. Participation is not open-ended, and the issue of *what is being participated in* is substantially externally decided. There is a strong emphasis on development on the basis of the non-consumption of natural capital, as opposed to the traditional notion of development as involving change to, and change in the relationship of people to, the natural environment. Even where outcomes cannot be argued to be subject to a conservation imperative (the cases of SNV, Tourism Concern and the IYE), the parameters of what can be participated in are constrained by the emphasis on natural capital in development, and the rejection of more thoroughgoing development – development that may go some way to changing a community's relationship with, and indeed reliance upon, their immediate natural environment – as not sustainable.

Fourth, there is generally a lack of consideration of the relationship between local level participation and development through ecotourism, and wider national development priorities. This is surprising, given that the nation state is the pre-eminent political and economic unit in the majority of social and political thought. Where this is alluded to, it is generally in the context of an argument for the benefits of a transfer of power from national to local levels of governance, the latter being the levels at which ICDPs operate. These features of the advocacy of community participation are critiqued in the following section, with reference to both the above case study summaries and the relevant academic literature.

Finally, community participation is presented as an 'alternative', and often as a radical agenda, based upon its neopopulist association with the agency of local communities. However, seen in context, it plays another role – that of limiting development possibilities to local, 'sustainable', externally decided agendas. As such, counter intuitively, community participation forecloses substantial development choices and narrows the scope for developing world societies to make their own future.

Facilitation and control . . . but of what?

The rhetoric of community participation, and the attendant concepts of empowerment and control, are very prominent in the case studies. The corollary of this is that the NGOs emphasise their role as facilitators – they bring to bear expertise and advice, and bring together the relevant stakeholders – but essentially are facilitating a process in which the community is central, rather than being the driving force in this process.

Brohman advocates this facilitation role, and sees it in terms of the devolving of 'political control' to the local level in tourism ICDPs (Brohman 1996b), a sentiment also prominent in the case studies. Yet the term 'political control' may be misleading. Control in the case of ecotourism ICDPs is linked to funding, and the funding is invariably tied to the outlook or the interest of the donor. In the case of the conservation-oriented case studies, this is clearly the case – there is an imperative to promote conservation regardless of the desires of the community. When the community has an alternative choice for its development, as in the case of the Innuit communities in northern Canada, conservation takes priority over community participation. Ultimately, the sites where ecotourism ICDPs will be funded by these organisations are determined by environmental considerations rather than human ones. This is, of course, unsurprising – these are, after all, organisations founded and developed on the basis of environmental preservation. The emphasis on community participation, however, presents the mission of these organisations as one of a relatively disinterested facilitator of the wishes of others.

In the case of SNV this is rather less clear. SNV, as principally a development organisation, potentially has a more open-ended agenda with regard to the shape of rural development. However, SNV's assistance emphasises the goal of achieving sustainable development based on the existing resources a community has, as shown in the summary. This is presented as a virtue, and is associated with sustainable development (SNV/Caalders and Cottrell 2001). Sustainable development, on the basis of one's natural capital, is likely to yield similar results to those of the conservation organisations in practice – if a community is dependent on its natural resources, then to conserve these resources may well, given such meagre development opportunities, provide limited benefits for the community. And as SNV's support is offered on this basis, rather than on an alternative, more open-ended one, offering the possibility of wider economic transformation, wider development is not an option.

It is the case that some projects have emerged from requests from the communities themselves, and hence the NGOs' claim to be facilitators may carry more weight here. In the interviews conducted, representatives from SNV, CI and Tourism Concern all cited examples where their involvement had been in response to requests either from communities directly, or from governmental authorities (Barnett 2000; Leijzer 2002; Sweeting 2002). However, even here the options for communities – the basis of their requests in the first place – are very constrained. The requests are likely to be a pragmatic response to the general dearth of development, an aid environment influenced by the 'greening of aid' and a recognition that ecotourism is an expanding sector. They do not amount to an endorsement of ICDPs as a development strategy.

Even if community involvement is as thorough as one could possibly imagine, this is only 'control' in a limited sense of the word. 'Political control' (Brohman 1996b), and a considerable amount of the rhetoric referred to in the summaries, implies that the *aims* of the funding, the trajectory of the development itself, is decided by the local population, and this is certainly not the case.

Empowerment and its attendant limits

As with 'facilitation' of the community, 'empowerment' has become ubiquitous as a justification for community participation, and is a prominent point of reference for the case studies and, as such, deserves attention. Thus far, we have taken empowerment to refer to the increased ability of individuals or communities to influence their destiny. Scheyvens provides a more detailed and nuanced definition of empowerment (1999: 247–9; see also Scheyvens 2002). She identifies four related aspects of empowerment that she believes should be features of ecotourism. These are listed below with brief definitions:

Economic empowerment

Lasting economic gains that are spread within the community.

Psychological empowerment

Relates to the self-esteem of members of the community, enhanced due to, for example, outside recognition of the 'uniqueness and value of their culture' and their 'traditional knowledge' (Scheyvens 1999: 247).

Social empowerment

Social empowerment is held to have been achieved when '[e]cotourism maintains or enhances the local community's equilibrium' (ibid.) and when '[c]ommunity cohesion is improved' (ibid.) through the project.

Political empowerment

> The community's political structure, which fairly represents the needs and interests of all community groups, provides a forum through which people can raise questions relating to the ecotourism venture and have their concerns dealt with. Agencies initiating or implementing the ecotourism venture seek out the opinions of community groups (including special interest groups of women, youths and other socially disadvantaged groups) and provide opportunities for them to be represented on decision making bodies.
>
> (ibid.)

What is notable about this categorisation is its limiting of the issue of power to the level of the community. All four of the categories, in so far as they might be regarded as political, or to do with the contestation of power, are micro-political categories – they pertain to politics *within* the community, in which the protagonists are individuals and interest groups. Even the widest category, that of 'political empowerment' (Scheyvens' definition of which is reproduced in full above) conceives of politics exclusively as internal to the community.

Elsewhere, Scheyvens also usefully provides critical comment on 'community' – the people to be empowered – that brings into the discussion different social networks and divisions within communities between rich and poor (Scheyvens 2002: 16). This writing exhibits a thorough consideration of community empowerment, but one that ultimately remains restricted to the internal dynamics of the community itself.

Scheyvens' definitions frame the outlook of the NGOs featured in the case studies. Here too, empowerment is restricted to implementing projects and distributing benefits – it is viewed as internal, within the communities. Even for Tourism Concern, which is at the forefront of campaigning for an 'ethical tourism' in which community participation is to the fore, this is the case.

In reality the parameters of empowerment are substantially given prior to the process itself. According to Akama:

> the local community need to be empowered to decide what forms of tourism facilities and wildlife conservation programmes they want to be developed in their respective communities, and how the tourism cost and benefits are to be shared amongst different stakeholders.
>
> (Akama 1996: 573)

However, that a mix of small-scale tourism and conservation is what is to be funded is beyond participation – it is established prior to the project itself, and, in the sort of projects Akama is referring to, funding will be conditional upon its acceptance. Akama criticises 'Western' environmental values, and argues that the community should be 'empowered' to overcome Western bias

(Akama 1996). But the empowerment invoked by Akama is largely illusory, its limits determined by a prior conception of what is desirable development.

In locating the issue of power within the community, empowerment eschews what might be regarded as a social understanding of power. A social understanding would inevitably consider the external relationship of the community to the world market, to Western aid agencies and to NGOs themselves. None of these seem to be features of empowerment as constituted by Scheyvens (2002). Indeed, her category 'psychological empowerment' is notable, too, in that it locates power at the level of an individual's psyche, which could be regarded as the antithesis of a social understanding. Broader issues of power between nations, between the developed and developing world, between social classes, and notions of social power beyond the immediate experience of individuals, are either deprioritised, or non-existent in this and other accounts that focus on empowerment. Reed (1997), for example, looks directly at the issue of power in her article entitled 'Power Relations and Community Based Tourism Planning', yet the conception of power adopted is restricted to interpersonal and inter-group power within the community. This is hardly surprising, given her definition of power as 'the ability to impose one's will or advance one's own interest' (ibid.). Such a subjective and general conception of power easily conflates social power with inter-group relationships, and in this case substitutes the latter for the former.

Elsewhere, research has been conducted, and critical accounts written, about how well various projects fare when their performance on participation is measured. Probably the most commonly invoked example of a scale for gauging this is Pretty's typology (Pretty 1995; see also Scheyvens 2002: 55). This typology can be read as a gauge of the thoroughness of empowerment as set out by Scheyvens, referred to above. Pretty's seven levels of participation feature 'manipulation' at one end and 'self mobilisation' at the other. Pretty's analysis presents a greater level of participation as 'good', with the ideal being this 'self mobilisation'. Here, communities instigate, as well as plan and see through, conservation and development projects within their community.

Yet Pretty's typology, too, emphasises the question of the *distribution of power within a community*. Its focus is on interpersonal and inter-group power. It has nothing to say about the prior limits placed on the community from without. If living a subsistence existence, closely reliant on the immediate natural environment, is considered a limitation on the community's ability to develop economically, then such limits are not challenged by ecotourism ICDPs. These projects tie development possibilities to the conservation of the immediate natural environment. Yet through the language of 'empowerment', 'participation' and 'control', such limits are presented as reflecting the agency of the community – *their* culture and *their* aspirations. The key issue arising here is whether the lauding of empowerment on a micro political level rationalises, or makes acceptable, a lack of power, or unequal power relations between the developed and developing worlds. If so, empowerment and community participation, as central aspects of the advocacy of ecotourism ICDPs, may be less than progressive.

As Mowforth and Munt point out, 'the push for local participation comes from a position of power, the first world' (Mowforth and Munt 1998: 242). Yet the community participation agenda that has become the focus of many people's aspirations to 'empower' developing world communities eschews these power relations between the developed and developing worlds in favour of the micro politics of the community. The extent to which power, control and democracy, and other related ideas invoked in the advocacy of ecotourism ICDPs, can be understood in this limited arena is questionable.

Democracy and control . . . or simple pragmatism?

Community participation also suggests itself as part of a democratic agenda – greater choice, empowerment and control all evoke a greater degree of democracy in development. The neopopulist tradition underpinning ecotourism has at its heart a promotion of the agency of the popular majority, usually within a locality. Tosun even asserts, with reference to tourism development, that without community participation, 'democracy and individual liberty may not be sustainable' (Tosun 2000: 615).

Yet a few writers have noted the obvious dilemma in community participation. What happens when communities opt for alternatives – mass tourism perhaps – that are not in keeping with the aims of funding authorities such as NGOs or Western development agencies? Weaver articulates this as follows: 'If [these] experts attempt to impose an AT (alternative tourism) model or to re-educate the local people so that they change their preferences, the entire issue of local decision making, control and community based tourism is called into question' (Weaver 1998: 15). However, this dilemma may rarely surface, as though communities may have many opportunities to engage with how a project is implemented, and how its benefits are distributed, the broader issue of choosing development priorities is foreclosed – there simply is not a mechanism through which communities can play a part in this. Where they do have an alternative – as in the case of the Innuit communities in the Arctic – they may well choose priorities deemed to be 'unsustainable' by some.

Jon Tinker, President of the Panos Institute, questions the democratic credentials of many aid projects, arguing that developing world communities are 'seduced by western NGOs into accepting their projects on their terms' (cited in Scheyvens 2002: 231). In reality, it may be less a case of seduction, and more one of pragmatism. Faced with the possibility of assistance tied to a particular type of project, or no assistance at all, the pragmatic choice is to accept assistance regardless of any unfavourable terms attached (White 2000). Hence, participation does not involve real choice at all, as there is an absence of alternatives on offer (ibid.). Rather, participation by the community is likely to be instrumental to the prospect of some limited financial assistance, the terms of which they have little if any control over (ibid.). Also, the language through which this funding is rationalised and presented by donors ('participation', 'ownership', 'empowerment' etc.) is likely to be adopted by recipients based on a recognition of its instrumental value rather than a deep-seated commitment to the development ideas it expresses (Hann and Dunn 1996). Hence, the democratic credentials of participation, seen in a slightly wider context, are illusory.

Joseph (2001: 148) argues that there is a danger that NGO projects based on 'participatory democracy', 'local development', 'citizen participation' and 'human rights' may 'in effect restrict participatory democracy and citizenship simply to participation at the micro level in processes and programmes to combat poverty and other effects of structural adjustment'. There is hence 'a tendency to regard any successful poverty relief programme at the micro level as "local development"'. This concept 'then loses any relation to envisaging and working towards other more holistic and more human forms of development'. NGO localism is, as he points out, a 'limited – often negative – concept of what politics is about' (ibid.: 149).

Joseph's analysis sheds light on the perverse nature of community participation in ecotourism ICDPs. It is only through democratic political activity that people can have some control over bigger political issues such as the trajectory of development. Localism situates control at the local level, and assumes that the big question of the type of development has been resolved beyond debate in favour of sustainable development, which in this case is interpreted as being rooted in the pre-existing relationship between people and their local environment. Certainly, as we have seen in the case of ecotourism, the national government is often ignored or denigrated. Agency is talked up at the local level, but at the same time limited to that level. Hence, the possibility to act beyond the immediate issue of managing local resources is ruled out of court in this discourse – it rarely features in analyses. The democratic credentials of ecotourism ICDPs are, then, at best illusory and at worst denigrate democracy by limiting it to the local level.

Did developing world communities choose the neopopulist approach to rural development?

Although it cannot be fully developed here, it is worth also considering the extent to which the rise of neopopulist 'development from below', and the consequent promotion of community participation, emerged from developing world societies themselves. If this were the case, it would add credibility to the view that community participation represents a democratic and empowering innovation – it would be, in a sense, 'of the people', something that emerged, or perhaps was demanded, from developing world societies themselves.

In fact, far from challenging development as modernisation with small-scale participatory alternatives, many post-colonial nations expressed the desire to mirror the industrialisation they saw in the developed world. According to two authors, many developing world nations

> took it for granted that western industrialised countries were already developed and that the cure for 'underdevelopment' was, accordingly, to become as much as possible like them. This seemed to suggest that the royal road to 'catching up' was through an accelerated process of urbanisation.
>
> (Friedman and Weaver 1979: 91)

So, while neopopulist community participation developed as an alternative to the experience of the developed countries, these authors note that many developing countries, having shed colonial status through the actions of truly popular national liberation movements, sought to join the 'modern', industrialised, urbanised developed world, rather than to retreat from its excesses by integrating conservation into development. This point is also developed by Adams (2001) and Preston (1996).

It is fairly easy to sustain the argument that communities will, and indeed have, as the organisations featured in the case studies argue, chosen to participate in ecotourism ICDPs and accrue the limited economic benefits from these. In general, since the 1980s there has been a marked change in thinking on conservation strategies towards factoring in the needs of local people, and unsurprisingly representatives from the developing world themselves have argued for this. The IUCN debate by its Specialist Group on Sustainable Use in 1991 is an example of this (Allen and Edwards 1995). Here, representatives of developing world countries countered the protectionist approach to wild animal conservation emanating from some developed world NGOs by advocating a conservation policy that took account of the dependence of rural communities on these wild species as economic resources. They argued that Western conservation should at least yield economic benefits for developing world people. A further example is the debates at the Third and Fourth World Congresses on National Parks in Protected Areas, held in Bali and Caracas respectively, at which representatives of indigenous groups put their case to be beneficiaries from conservation policies, rather than excluded through a 'fences and fines' approach (Wells and Brandon 1992).

Yet to infer from this that the general conception of development, or of sustainable development, underlying the choices presented to rural developing world communities has a legitimacy borne of participation is surely a false inference. The rhetoric surrounding participation emanating from the case studies, and also from neopopulist advocates of ecotourism, serves to present the emphasis on conservation in development thinking as a product of the agency of the communities, with all the legitimacy and credibility that flows from this. It would be more accurate to say that the community participates in the implementation of projects shaped elsewhere.

Critics of community participation

It would be wrong to argue that community participation, and its associated armoury of terms such as empowerment and control, are accepted uncritically in the literature – or indeed by the NGOs in the case studies. For example, Midgeley writes with insight that 'the notion of community participation is deeply ideological in that it reflects beliefs derived from social and political theories about how societies should be organised' (Midgeley 1986: 4). Midgeley is referring here to the notion that the rhetoric of community participation could be a cover for Western-style 'modernisation', an argument also prominent in Mowforth and Munt's *Tourism and Sustainability: New Tourism in the Third World* (1998).

In fact, as noted earlier, many critics have questioned the efficacy of community participation along these lines, regarding it as either tokenistic, or a cover for development or preservationist schemas emanating from funders (Cooke and Kothari 2001). For example, Woodwood's research argued that the norm in South African ecotourism projects was to adopt a participatory approach primarily in terms of its public relations value (Woodwood 1997: 166). Similarly, Scheyvens cites the work of the Conservation Corporation of Africa (CCA) as an example of an organisation that she believes works with local communities only out of a sense of economic pragmatism rather than a commitment to the communities themselves (Scheyvens 2002: 192–3). CCA is a private company, not an NGO. Scheyvens quotes and reproaches the Phinda reserve manager, who acknowledges that the compliance of the community is on the basis of: 'If they poach, it's not us they're stealing from but themselves', rather than a philosophical commitment to greater democracy and equity (ibid.).

Yet is this approach, roundly criticised by Scheyvens, so different from the alternative examples she and others cite as being progressive? Private companies may introduce participation on an instrumental basis, for its public relations value. In the case of the conservation organisations, there seems to be a similar instrumental approach to participation – it is participation *for a specific end*, an end no more the product of the community's unfettered desires than in the case of the CCA in the above example. And Midgeley's statement, referred to above, that community participation is 'deeply ideological' (Midgeley 1986: 4) holds true with regard to the neopopulist alternatives too – these alternatives have been developed in, and are funded from, a particular milieu in the developed world. They, too, are ideological – that they emanate from civil society, rather than government or commerce, does not preclude this.

Within the case studies, too, there is criticism and self-criticism along similar lines. Ecotourism ICDPs, as a recent innovation, have an experimental character, and it is widely accepted that there is room for improvement. Tourism Concern consistently takes to task NGOs for their lack of meaningful community participation, and the resultant failure of communities to benefit adequately from projects (e.g. Tourism Concern 1995). Yet the conservation NGOs, WWF and CI, stress community participation in their literature, and are critical of their own conservation policies of the past, policies regarded by some as 'fortress conservation'. All appear implicitly aware of the criticisms of community participation as tokenism, and keen to place it at the centre of their work, perhaps in response to these criticisms.

Yet while community participation remains such a central, and contested, focus of ecotourism, what all the case studies and the vast majority of the literature all share is a support for the broader, underlying project of integrating conservation with development in impoverished rural communities, and hence a focus on the non-consumption of natural capital as the basis for economic progress. This 'bottom line' is associated with sustainable development, and is not substantially challenged or interrogated within the case studies or the associated general literature. Thus, that community participation itself could be a conduit for an imposed agenda is overlooked. It is conceivable that this may have something to do with the moral force of the invocation of 'community' and 'sustainable development' (Butcher 2003a), and also of the apparently agenda-less civil society roots of the NGOs (Kumar 1993).

Local participation and national priorities – an unexamined tension?

Towards the start of the chapter, the emphasis on the community as a local, as opposed to national, phenomenon, as the appropriate spatial unit for development, was established as a feature of neopopulist thinking. There is relatively little justification for this within the tourism-specific literature – writing on national tourism trends on one hand, and on rural, participatory tourism on the other, seldom meet in the middle.

Yet the idea of community can equally be applied to the nation – the national community – and inevitably what takes place, or does not take place, in the forests of a developing world country, also affects people living in the cities. It is not clear where this leaves local participation in relation to national participation through elections – should the local be privileged over the national, as it seems to be in the thinking behind ecotourism ICDPs?

One author who does consider critically the relationship between community participation in ecotourism and national priorities, Scheyvens (2002), simply sees the issue in terms of central government's role in facilitating community-level development. The national strategies she advocates are 'an appropriate policy environment, regulatory framework, infrastructure and support for small business development' and to 'give priority to investors working to assist local communities, and grant communities secure tenure over their land and other resources' (ibid.: 244). Here, the national policy is about backing up the community, and even ceding control of land to localities. Scheyvens' view reduces national priorities to acting as an enabling state in rural areas, enabling the functioning of locally based, sustainable development. She is ultimately interested in development through 'local agency' (ibid.: 56). This is a common theme, developed by Chambers (1993: 121) and other neopopulist writers. It is not at all clear where this leaves national agency and the imperative of national development, in a global world economy based upon trade between nation states.

In similar vein, Parnwell argues that community participation is desirable in order to compensate for a lack of democracy or good governance at a national level. He argues that the ability of NGOs and communities themselves to shape tourism in a fashion that is positive for the community, depends on the 'prevailing socio-economic context' (Parnwell 1998: 217), and goes on to contend that developing world governments may encourage international capital to benefit the elites rather than to benefit the majority of the people. Once again, the state is represented as a limiting factor upon the community, and the ability of the community, in conjunction with NGOs, to develop and conserve through ecotourism.

It is evident in the case study summaries that the view espoused by Scheyvens and Parnwell – a perspective central to neopopulist thought – is shared by the featured NGOs, especially the cases of the IYE and SNV. In addition, the call to grant tenure over indigenous land has been heeded by WWF and SNV, which have successfully argued for this in Zimbabwe and Botswana respectively.

This privileging of the local community over the national government does have merit. For example, Scheyvens is partly right to claim that modernist discourse has been preoccupied with macro level improvements, rather than a broader concern for well-being (2002: 33). However, the critics seem to morally elevate the local, community level above macro level indicators, and indeed quite a lot of the discussion about development through ecotourism fails to mention national perspectives at all. As such, they may replace a bias toward macro indicators – a national bias which neopopulists claim is symptomatic of modernisation as development – with an inability to envisage development as anything other than a locally based phenomenon.

For example, Scheyvens (2002: 54) articulates the case for ecodevelopment through ecotourism, arguing that 'A concern for livelihoods should be integral to development efforts, based on the recognition that local people need to benefit from the existence of natural resources in their area . . .' To argue that a concern for livelihoods should be central to development is uncontentious, but equally vague. However, to suggest that *local* people need to benefit from the existence of natural resources *in their area* is more difficult to accept. In most contexts in developed countries, with an international division of labour and global trade, people do not benefit from the natural resources in their area. They tend to benefit from resources in the widest sense – every time they switch on a light, light the gas oven, drive their car, read a book or visit a museum they are benefiting from resources produced far from their own communities. Resources that communities in the developed world have at their disposal, and the efficiency with which they can transform them into goods, is shaped by modern development which itself is premised on an international division of labour. Scheyvens' localised, ecodevelopment approach eschews this legacy in favour of self-sufficiency and smallness of scale, not as a stepping stone to something else, but as a point of principle.

Also, concerns about the lack of a 'trickle down effect' from nationally based development can be well justified (e.g. Scheyvens 2002: 8; Hitchcock *et al.* 1993; Butler 1990). However, these concerns seem to dismiss rather readily that in a world of nation states, development on any substantial scale has to have a strong national perspective if it is to contribute to the transformation of national economies away from developing world status towards a more developed one.

Among the benefits of ecotourism projects, their advocates argue, is their ability to link with the local, informal sector, such as the production of crafts, thus ensuring that the poor see direct benefits (Opperman 1993). Yet if the ability to link with these informal economic circuits is a strength, it is also a weakness. Informal circuits may alleviate poverty locally, but economic development requires the development of the formal economy (which is also the tax paying economy), feeding into national development. Although local, informal linkages are often talked up in the advocacy of ecotourism, their ability to contribute to significant economic development should be questioned. Here again, the 'advantages' of localism can only be sustained if we accept that rural development is tied to the pre-existing relationship of small communities to their surrounding natural environment.

Advocates of community-based development often argue that development should be not what is done to people, by states, bankers and 'experts', but rather it should be what human communities do to themselves – it should be endogenous. This formulation is, on the face of it, almost impossible to disagree with. However, in the case study summaries the community is invariably a local one. Yet this does beg the question, cannot the community be recognised as a national community, and the state the most legitimate representative of that community? Also, it is self-evident that, globalisation notwithstanding, we live in a world in which nation states are the principal unit of economy and politics . Surely, planning and key decisions on the use of scarce aid should logically have a very strong national component. Infrastructural development, for example, is almost impossible to conceive of when focusing on local level development – by its very nature much infrastructure links communities to resources, other communities, the nation and beyond.

Community participation – a radical agenda?

It is notable that community participation is often viewed by its NGO advocates as radical, as a counter to overbearing governments and the rhetorical free-market agenda associated with the big global financial institutions the World Bank and International Monetary Fund (IMF). This is the tenor of Scheyvens' *Tourism and Empowerment* (2002) and of Tourism Concern's *Community Tourism Guide* (Tourism Concern/Mann 2000). Activist Anita Pleumaron even argues that true 'grassroots' participation is necessary as part of the construction of 'an alternative "new world order" in which people themselves, rather than outside interests, determine and control their lives' (Pleumaron 1994: 147). Yet the radical rhetoric of 'empowerment' and 'community control' masks a shared outlook with the proponents of the free-market rhetoric that the advocates of community participation often claim to oppose, that shared agenda being a diminished view of the importance and efficacy of the state in development.

This is clear if we compare the New Policy Agenda (the term sometimes given to the 'New Right' emphasis on markets in development, especially in the 1980s and subsequently), and the Alternative Development Paradigm (the alternative, 'people' oriented view of many NGOs, often associated with the Left, and consistently associated with the promotion of community participation). The Alternative Development Paradigm, situated in the cultural and environmental 'Left', has increasingly turned away from the state, associating it with failed grand development schemas, and has adopted a neopopulist localism as a key priority. One author is frank enough to admit that 'putting people in the centre of development implied removing the state and its agents from that centre' (Tandon 2001: 53). From the perspective of the New Policy Agenda, the developing world state was an inefficient and bureaucratic burden upon business, and needed to slim down and adopt a set of free-market oriented policies (a view developing world states were impelled to take on board in order to benefit from debt relief under structural adjustment policies). Hence, the shared assumption between these two apparently contrary viewpoints is a diminished view of the role of the state and sovereignty (Feldman 1997).

The role of the developing world state in social development has been curtailed by debt, fiscal difficulties and the imposition of structural adjustment (ibid.; Midgeley 2003), and economic and social policies have progressively become less the concern of sovereign governments, and more the product of the will of global financial institutions, notably the World Bank and IMF. Yet the role of NGOs has reinforced rather than challenged this state of affairs. As the capacity of states to intervene in their own societies has reduced, the space for the growing role of external agencies has opened up. Since the 1980s, as Powell and Seddon (1997: 10) argue, the aid industry, via NGOs, has exerted control over ever more detail of the development agenda and introduced itself more powerfully into civil society as an alternative to the state (ibid.). In this light, there is a logical case to be made that ecotourism ICDPs and similar interventions may undercut the authority of the state, effectively establishing an alternative focus for communities seeking assistance. Bebbington and Riddell make a general point that has passed by the NGO oriented advocates of ecotourism – that 'moves by donors to support NGOs rather than government merely weaken government further. The argument that NGOs are a better alternative then becomes little more than a self fulfilling prophecy' (cited in Hulme and Edwards 1996: 114).

NGOs can be seen as the conscience of the New Policy Agenda – purporting to deal with the environmental effects of modern development and help the poorest living on the margins of the economy. In many instances NGOs have become, in the eyes of funders, a 'favoured child', and the 'preferred channel for service provision *in deliberate substitution for the state*' (Edwards and Hulme 1995: 4–5; my italics). Hence, in an important sense they have become implementers of the New Policy Agenda (Hulme and Edwards 1996: 114), or are perhaps simply dealing with the effects of the neo-liberal model, rather than constituting an alternative (Joseph 2001: 150–1). Yet at the same time their stance reinforces the denial of sovereignty implicit in structural adjustment. In the case of ecotourism ICDPs, there is also a distinct emphasis on self help, entrepreneurship and small business, a rhetoric that fits well with the rhetoric of the Right.

The apparent distance between but substantial congruence of the two positions – Alternative Development Paradigm and New Policy Agenda – suggests that the issues raised cannot be readily understood in terms of the politics of Left and Right at all. Both NGO radicals and free-market advocates of the NPA in Washington share a dim view of the developing world state and its sovereignty in its economic affairs. National sovereignty has been viewed in the past in the liberal humanist tradition as an articulation of the freedom of peoples in the face of domination – the right of nations to self-determination was a principle championed (if not always practised) by revolutions in America (1775–83), France (1789), Europe (1848) and Russia (1917). In the post-Second World War period, sovereignty was fought for, won and celebrated in many former colonies that comprise the developing world. The fight for sovereignty was seen as a demand for national freedom, an expression of the political agency of a people. Yet today, a defence of sovereignty finds few allies from Left or Right, from global free marketeer to radical environmentally or culturally oriented NGO activist. Instead, the issue of agency is restated as local empowerment via NGO initiatives such as ICDPs, a poor substitute for national sovereignty.

Although this cannot be developed here, it is worth reminding ourselves that the balance sheet of achievements of African states in the post colonial period up to the 1980s was generally far better in terms of extending education, healthcare and basic services, than has been the case in the more recent era in which the Alternative Development Paradigm (and the New Policy Agenda) has been influential. Of course, such a comparison is quite limited – the geopolitical outlook for the developing world was very different in the years following colonialism, and during the cold war, than in subsequent decades. Yet it reminds us that NGO interventions are a poor substitute for national development plans of sovereign states. At most, NGOs are assigned to deal with the consequences of the lack of general development prospects, focusing on basic needs and, in the case of ICDPs, maintaining a steady state between marginalised rural communities and their environment.

This parallel between these two apparently opposing outlooks – New Policy Agenda and Alternative Development Paradigm – is also evident if we consider further structural adjustment policies associated with the World Bank and IMF, and compare them to the debt for nature swaps initiated by environmental NGOs (including CI and WWF), and often involving ecotourism projects. These two policies – from the perspective of the advocates of the New Policy Agenda and the Alternative Development Paradigm respectively – are illustrative of a common diminution of state sovereignty.

Structural adjustment programmes (SAPs) have been widely utilised by the international financial institutions to dictate economic reforms to developing world states, using debt as leverage. As such, it was central to the New Policy Agenda. Simply, more favourable terms for debt become obtainable on condition that economies liberalise and reduce the state's role. Through SAPs the World Bank and IMF 'virtually control the economies' of many developing countries (Potter *et al.* 1999: 169). This clearly calls into question state sovereignty. Structural adjustment programmes were and are prominent in undermining any room to manoeuvre that indebted states may have had in policy, and served to undermine public services such as education, health and other programmes of public works.

Debt for nature swaps work on a similar principle – wealthy environmental NGOs offer to buy up a portion of debt and reduce its burden on the society in question through rescheduling onto a more favourable basis, in return for effective control over the use (or non-use) of swathes of land they consider important for conservation. The aim may be very different, but the implications for sovereignty are similar – debt is used as leverage to impose an externally decided priority. This diminution of sovereignty implicit in debt for nature swaps is in keeping with the neopopulist advocacy of ICDPs generally – it is not an aberration, but a logical development of the ecocentric thinking underpinning them. This thinking holds that local communities should subsist in a 'sustainable' relationship with their local resources, in such a way as to promote the provision of basic needs, but with little prospect of going beyond this – a scenario glibly labelled 'sustainable development' in many texts and papers. In such formulations, national planning and national interests become a burden on the local community and environment. This sentiment runs through the case study summaries (notably the IYE) and is influential in the wider literature (notably Scheyvens 2002).

The talking up of the community – always a local community, never a national, and rarely a regional one – is accompanied by a denigration of the nation's ability to achieve progress for its people. The lack of democracy, and poor governance generally, may provide pragmatic arguments for NGOs to operate at a local level. However, ultimately an important but rarely asked question remains; how far can local community-based projects contribute to any sort of transformation of the economic prospects of a nation, and through doing so, increase the ability of the centre to govern? The denigration of the developing world state is based on real, not just ideological, factors – states are often affected by corruption, inefficiency and a lack of legitimacy. But the problem is that the promotion of local development, linking in to local needs on a small scale, does not address this problem – rather, it simply turns away from it. As a pragmatic means of relieving poverty in specific instances, the local emphasis of ecotourism ICDPs may have merits. However, the promotion of the local, and the denigration of the national, is part of a particular development philosophy, that promotes itself not as pragmatism in the face of inadequate aid budgets and poor governance, but as 'sustainable development'.

Conclusion

It is a truism that, in any given circumstance, it would seem to be better to seek out the views of those affected by development, even if this results in only minimal change to the development project itself.

However, the claims made for community participation go a lot further than this. Community participation in ecotourism is presented as an ethical approach to development, running counter to previous forms of development that did not seek to involve the community. It is presented as having the potential to substantially shift power over development to the communities themselves. Formally, community participation may be very thorough. However, the extent of choice over what is being participated in is very limited. It would seem that participation is instrumental – it acts as a means to organise and involve, and to give people a stake in projects. Ultimately, community participation is about negotiating the terms on which a project is to be implemented, rather than about the nature of the development project itself. To engage with this may simply be the pragmatic option for communities, given that available aid funding is linked to the acceptance of these projects.

As such, 'control', 'empowerment' and 'democracy' need to be tempered by a recognition that community participation on the part of the NGOs is intrinsic to a particular development agenda, an agenda shaped externally, and presented to poor rural communities as their sole option, this justified through the language of sustainability. In so far as it serves to legitimise that agenda, by attaching democratic credentials to it, it could be criticised as contributing to a limiting of development options through a narrowing of the development agenda to that which is local, small scale and 'sustainable'.

This chapter has examined the character of the rhetoric of community participation emanating from the case studies. One feature of this, as noted above, is that the community is invariably conceived of at a local level – ecotourism ICDPs claim to empower the local community through giving them a greater say and a greater stake in development. Chapters 5 and 6 consider further what it is, precisely, that the communities are offered a stake in. They will consider the extent to which the conception of development is bounded by a particular reading of culture and cultural change emphasising tradition, and by a particular conception of environmental fragility that insists upon the non-consumption of natural capital.

5 Tradition in the advocacy of ecotourism

Introduction

Ecotourism, as marketed to tourists, is normally linked to 'traditional' culture at the destination. The invocation of 'traditional ways of life', 'local customs' and 'authentic culture' are common refrains in the advertising of this growing niche market (Butcher 2003a). The 'new tourist' (Poon 1993) desires to witness examples of cultures different from their own, perhaps centring themselves spiritually (Cohen 1972: 165) in developing world cultures that in some way exhibit a closer relationship between people and the natural world (Fennell 2003; Krippendorf 1987). Indeed, it can be argued that the growth of ecotourism itself is in part a response to a profound disillusionment in the developed world with the experience and the outcomes of development (Butcher 2003a).

This emphasis on tradition is mirrored in the discussion of the development merits of ecotourism. There is a strong emphasis on tradition in the conceptualisations of culture and cultural change in its advocacy. Specifically, the preservation of traditional knowledge is deemed central to sustainable tourism development.

This chapter reviews the pertinent documentary literature for each case study in turn. Following a critical review of each case study, common and divergent themes are identified, and the chapter offers an analysis of the significance of the emphasis on traditional culture based on this.

A number of important themes arising from the discourse are considered: the notions of culture entrenched in the past; cultural relativism; culture as functional; and also the apparent contradiction in the support for local traditions through external intervention.

Specifically, it is argued that the functional and relativistic approach to culture implicitly and explicitly adopted in the advocacy of ecotourism as sustainable development leads to a restricted and restrictive conceptualisation of development.

The chapter also comments on the claim that traditions in rural communities embody a way of thinking that is benign towards the environment, sometimes referred to as 'the environmentalism of the poor' – a notion strongly implied in the advocacy of ecotourism. Finally, it is suggested that far from the emphasis on tradition reflecting the culture of the community, it would be more accurate to say that it is made in the West.

A note on terminology – tradition, traditional societies and traditional knowledge

Traditional societies are those that function on the basis of subsistence agriculture and are not substantially integrated into the world economy through trade and the international division of labour. They are societies or communities that have not been party to the process of modernisation associated with the developed world. They are hence sometimes referred to as marginal, and are inevitably economically poor. In such societies, traditions are rooted in a direct relationship to the natural environment, relatively unmediated through modern technology and the global division of labour.

Traditional knowledge (often presented as a counter to modern science by critics of modern development) refers to a cumulative body of knowledge and beliefs handed down through generations, about the relationship of people to their environment and to one another. Such knowledge is an attribute of traditional societies, societies with historical continuity in resource use and in their way of life generally.

Closely associated with this is the term 'indigenous knowledge'. In similar vein, indigenous knowledge has been defined as a body of knowledge indigenous people have accumulated over time, which allows them to live in balance with their environment. It is also sometimes considered to be an applied science as it is generated and transformed through a systematic process of observation, experimentation and adaptation.

Another source argues that knowledge can be described as indigenous if 'it originates from or is bound to local experiences, and takes its local world not perhaps as the only one in existence, but as being locally the most relevant of all' (Seeland 2000: 7). Indigenous peoples, or 'first peoples', are often discussed in relation to ecotourism projects. Their claim to the land precedes modernisation, and hence their traditions have a certain claim to authenticity in the advocacy of ecotourism (Johnston 2005).

A typical view of the importance of indigenous or traditional knowledge is that of the Mountain Partnership, a group consisting of funding agencies and NGOs (including WWF), set up at the World Summit for Sustainable Development in Johannesburg in 2002: 'Traditional knowledge and values are generally being supplanted by modern practices and values. But there is growing interest in capturing indigenous wisdom and applying it in development for better grassroots participation, improved sustainability and environmental conservation' (Mountain Partnership undated).

The association between traditional knowledge, participation and conservation is commonplace, and this association is evident in the case study summaries included in this chapter.

So, although such societies are inevitably influenced by the ether of global capitalism, they retain the characteristics of pre-modern societies. Often traditions revolve around the rhythms of nature, rhythms that impose themselves on the lives of rural developing world communities whose existence is closely linked to their immediate natural resources. Traditional or indigenous knowledge is the knowledge that arises from this relationship, such as that pertaining to traditional farming methods, traditional healing, distinctive building techniques using locally obtained materials, and small-scale craft production. Clearly, traditional knowledge is an important aspect of culture, especially in rural societies in the developing world that have not undergone the social change that accompanies modern development.

On the basis of the case studies and other literature advocating ecotourism it is clear that there is a strong emphasis on the role of local tradition and local knowledge in the development in these projects. Ecotourism appeals to the neopopulist ambition of drawing on the agency of the community, and the invocation of tradition, traditional knowledge, indigenous knowledge and other similar terms seems a worthy aim in this respect.

Interest in the role of traditional knowledge in development is not new. Allan's *The African Husbandman* (1965) recognised that indigenous agricultural systems demonstrate an important resource of knowledge of the environment, and marked an early interest in the subject (Adams 2001: 338). Serious interest in indigenous knowledge can be traced back to the 1980s (Briggs 2005: 100). However, traditional knowledge has really come to the fore parallel to the rise of sustainable development. Notably, Principle 22 of the Rio Declaration on Environment and Development argues that indigenous people and other local communities have a vital role in environmental management and development because of their knowledge and traditional practices. States, it is argued, should recognise and duly support their identity, culture and interests and enable their effective participation in the achievement of sustainable development (UN 1993).

In this spirit, traditional knowledge and culture generally are increasingly factored into the debate, and this is held to represent 'a shift from the preoccupation with the centralised, technically oriented solutions of the past decades' (Agrawal 1995: 414). Rich concurs that 'decades of failed international development projects' have 'ignore[d] and often destroy the local knowledge and social organisation on which sound stewardship of ecosystems as well as equitable economic development depend' (1994: 273). In response to this, the importance of traditional knowledge is codified at the level of rural development NGOs such as SNV, and conservation NGOs such as WWF and CI. The institutions of global governance of finance have also, rhetorically at least, taken it on board (e.g. World Bank 1998).

Yet, according to one critical author, the use of indigenous knowledge has, since the 1980s, 'become a kind of mantra . . . representing one possible way of negotiating the so called "development impasse" or indeed, the "death of development"' (Briggs 2005: 99), suggesting its influence comes more from the discrediting of the alternatives and less from a convincing case for the role of tradition in development.

The mantra of traditional knowledge is certainly in evidence in the advocacy of ecotourism. Indeed, the specific debate about ecotourism tends to amplify tradition. As well as reflecting the way of life of the society in question, tradition is also part of the attraction for prospective ecotourists who, disillusioned with the modern world, seek a taste of a simpler relationship to the natural world. Therefore, aspects of traditional knowledge and culture are very directly associated with development – the former are saleable, either as material culture (e.g. crafts) or as part of the experience sought by tourists (e.g. witnessing cultural rituals). Notably, some ecotourism projects very directly promote traditional knowledge as exemplary of a more environmentally friendly, sustainable way of living to ecotourists, as a way of life people in the developed world can learn a great deal from. A striking example of this is the Ladakh Farm project in India, funded by the International Society for Ecology and Culture. This project attempts to contribute to 'sustainable development' by raising the status of traditional, subsistence agricultural methods. It is also notable that this project sees traditional knowledge as an example for the richer nations thus:

> Travel [to Ladakh] can mean a lot more than a leisure activity. It might form part of a broader philosophical reflection relating to the self and nature. It might involve trying to find answers to many of the problems experienced when living in a westernised, industrialised country.
>
> (Acott *et al.* 1998: 240)

This example is not untypical of the deference to traditional knowledge in much of the advocacy of ecotourism – it is widely argued in the academic literature that an emphasis on traditional knowledge is not just a key aspect of the ecotourism product, but that this is a normative goal with regard to development (Fennell 2003; Wearing and Neil 1999; Johnston 2005).

The case studies and the issue of tradition

This section draws on the documentary evidence in the case studies concerning their approach to tradition in the advocacy of ecotourism as sustainable development.

WWF

In the WWF's literature on ecotourism there is certainly a great emphasis on support for tradition and traditional knowledge. One document argues that:

> [l]ocal traditions should be taken into account in buildings, and architectural development should be in harmony with the environment

> and the landscape. The knowledge and experience of local communities in sustainable resource management can make a major contribution to responsible tourism.
>
> (WWF-International 2001a: 3)

The document goes on to argue that '[t]ourism should therefore respect and value local knowledge and experience, maximise benefits to communities, and recruit, train, and employ local people at all levels' (ibid.).

This emphasis on a recognition of and sensitivity towards local traditions, linked here also to local economic benefits, is not generally regarded as contentious (although it does reflect the assumption of the local community as being the unit at which development should be conceived and organised, as discussed in Chapter 4).

However, WWF goes further than just advocating sensitivity to local knowledge. One document holds that 'products developed *should be based* on the communities' traditional knowledge, values and skills' (WWF-International/Denman 2001; my italics). Further, ecotourism should ensure that '[t]raditional styles and locally available materials *should* be used' (ibid.: 20; my italics). The argument continues that in some communities this could support thatchers, and existing buildings could be utilised to forgo the need for new developments. The report goes on to outline how projects' impacts can be kept as low as possible.

Here, development is to be based around traditional knowledge – the skills, crafts and other products connected to the way the community functions and has functioned in the past. Such an approach is posited as having positive potential for local producers, forgoing the need for new capital, keeping the environmental impact low and respecting the host community's culture (ibid.).

A link between the maintenance of tradition and sustainable development is also drawn by WWF – it argues that, in contrast to mainstream tourism development, ecotourism can 'enhance cultural and historical traditions which contribute to conservation and sustainable management of natural resources' (WWF-International 2001c: 2). Similar assertions are made elsewhere too (e.g. WWF-International 2001a: 3).

However, there can be contradictions in the advocacy of the preservation of tradition when this conflicts with the conservation priorities of WWF. For example, the trade in bush meat in Cameroon, the Central African Republic, the Republic of the Congo, the Democratic Republic of the Congo, Equatorial Guinea and Gabon has conflicted with the view taken by the CITES bush meat working group at its first meeting held in Cameroon in 2001 (WWF-UK undated d). Ecotourism is discussed as one possible way of changing the way local populations derive a living from the wildlife (ibid.). This clearly involves moulding tradition around externally decided priorities, rather than a fêting of tradition per se. Hence, in this case it seems that when tradition does not match the pre-existing priorities of the project, a modification of tradition may be advocated on the basis of sustainable use of resources.

WWF supports and works closely with an organisation called Terralingua, whose aim is to support 'biocultural diversity', defined as 'the integrated protection, maintenance and restoration of the world's biological, cultural and linguistic diversity' (Terralingua undated). The stance on biocultural diversity is an important point of reference here. While environmentalists have long made the case for preserving biodiversity, *biocultural* diversity ties culture into this project. The argument put forward by Terralingua is worth citing at length:

> Language, knowledge, and the environment have been intimately related throughout human history. This relationship is still apparent especially in indigenous, minority, and local societies that maintain close material and spiritual ties with their environments. Over generations, these peoples have accumulated a wealth of wisdom about their environments and its functions, management, and sustainable use. Traditional ecological knowledge and practices often make indigenous peoples, minorities, and local communities highly skilled and respectful stewards of the ecosystems in greatest need of protection. Local minority and indigenous languages are repositories and means of transmission of this knowledge and the related social behaviours, practices, and innovations.
>
> (Terralingua undated)

The argument explicit here is implicit in WWF's advocacy of ecotourism, and that of the other case studies. It holds that biodiversity is best preserved through the preservation of traditional aspects of culture, which are themselves diverse, reflecting the particular relationship between each community and the land, a point developed and made explicit in joint publications between WWF and Terralingua (Maffi and Oviedo 2000), and in a joint geographic information system (GIS) mapping project of the world's biocultural diversity (ibid.).

CI

CI views its mission in its work with ecotourism in the following terms: 'Our mission is to conserve the Earth's living heritage, our global biodiversity, and to demonstrate that human societies are able to live harmoniously with nature' (CI undated a: 1). This statement is exemplary of CI's outlook with regard to the culture of the host community, and to cultural change. 'Living heritage' is closely connected to biodiversity – it is cultural diversity that supports biodiversity, and vice versa. Such a relationship is regarded as 'harmonious' (ibid.).

A good example of CI's perspective on tradition in ecotourism is found in its 'World Legacy Awards', awarded by its Ecotravel Center, established to reward what it sees as the best practice in ecotourism development:

> The Ecotravel Center aims to showcase life affirming examples of sustainable, responsible tourism as a local economic alternative to more

> detrimental industries or practices of tourism. The awards will honour ecological leaders in the tourism industry that emphasise meaningful tourist experiences with the *current culture and heritage of people* and the diversity of nature at a particular destination.
>
> (CI undated e; my italics)

This 'current culture and heritage' is to be supported and sustained through ecotourism revenues: 'Placing economic values on a locale's natural, historical and cultural attributes strengthens an incentive to protect them. By sustaining nature and culture, tourism can survive as a viable economic activity for communities that inhabit the Earth's remaining richest environments' (CI undated e).

There is an emphasis in CI's thinking here, and elsewhere, on an intrinsic link between unique traditional cultures and unique environments. Indeed, it argues that places 'where unique human culture and natural ecosystems live should inspire the growth of sustainable tourism' (CI undated e).

Further, the role of ecotourism is to support a *harmony* between the community's way of life and its natural environment by ensuring that harmful impacts, impacts that would upset this harmony, are minimised. This is discussed in one document in terms of 'ecologically sound, harmonious encounters with people and nature that minimise harmful impacts on cultures and biodiversity in and around tourism sites' (CI undated e). There is, in this vein, a general sense that, in so far as development transforms culture, this is harmful.

Overall CI, as with WWF, draws a connection between traditional ways of life and conservation of the environment in a fashion essentially similar to the view of Terralingua and WWF in their joint work on biocultural diversity. Also, CI's ecotourism projects incentivise conservation on the basis of drawing on tradition as an economic asset. The adjective 'sustainable' is liberally applied to this relationship.

SNV

SNV's literature also places great stress on traditional knowledge. For example, commenting on CBT in the Xai-Xai tribal community in Tanzania, SNV advocates CBT on the basis of its role in providing 'a source of *cultural preservation* for the Bushmen' and 'an income generating project that is based on *knowledge that people already possess*' that is 'owned and directed by the people themselves' (SNV/Rozenmeijer 2001: 26; my italics).

Among the socio-cultural benefits of ecotourism is 'preservation of the cultural heritage of an area which might otherwise be lost as the result of general development' (SNV/Caalders and Cottrell 2001: 44). Here, limited development based upon cultural preservation is explicitly presented as better than other forms of development that may involve change to culture. Basing development around traditional knowledge in this way is striking when we consider the longstanding association of development as transformative of tradition, and based on modern scientific knowledge and technology.

There is a predisposition towards traditional culture in SNV's advocacy of ecotourism. There is certainly a sense in which the goal of preserving traditional culture is decided on prior to consultation with the community – in the literature it is simply assumed that this is desirable and is a goal shared by the community. For example, in discussing consultation, one SNV document asserts: 'It will be discussed in the village how commercialisation of the culture can be prevented' (SNV/de Jong 1999: 18). The issue to be discussed is 'how', not 'whether' this should happen.

Indeed, some of the projects appear to advocate an approach to traditional knowledge beyond sustainable development and into the realm of social engineering. For example, SNV asserts that CBT can promote the 're-valuation of individual cultural identity among the local population' (SNV/Caalders and Cottrell 2001: 11), and can lead to a 'reinforcement of a sense of pride by local people in their culture' (SNV/Caalders and Cottrell 2001: 44). In this vein, once '[t]raditional knowledge has become an economic asset for the project, elders in the community have enhanced their status because of their knowledge' (SNV/Rozenmeijer 2001: 41). Hence, the projects can reaffirm the waning status of traditional leaders in impoverished rural western Botswana.

One positive impact from SNV's projects is deemed to be the '*re-valuation of ecological values by the local population and authorities* as a result of tourism interest, as well as economic justification and means for protection of nature', and also that 'tourism may be less damaging to nature compared to alternative economic sectors such as agriculture and forestry (deforestation)' (SNV 2001: 11; my italics).

Traditional knowledge is also strongly associated with sustainable development by SNV. What it refers to as CBT 'makes use of traditional knowledge systems' and, through this, it 'can be an effective and sustainable way of making use of available natural resources in western Botswana' (SNV/Rozenmeijer 2001: 18). Further, in the same document, it is argued that 'this approach ensures a sustainable use of the resources it is based upon', these resources being the culture and the natural environment (ibid.).

Tradition becomes an economic resource through ecotourism. Traditional knowledge is often a central aspect of what the tourist desires to experience (Poon 1993), and they are prepared to pay for this – hence, it is deemed development on the basis of tradition, rather than destructive of it.

One example of the perceived role of traditional knowledge in attracting ecotourists is 'healing' based on ancient knowledge and mysticism (SNV/de Jong 1999: 6). 'Healing' is attractive to wealthy tourists, fascinated by the traditions and passionate about gaining a spiritual dimension to their lives. It is part of traditional knowledge, and, as such, reflects the pre-existing culture of the community. So in a sense 'healing' as an economic asset reinforces 'healing' as a cultural practice. Yet modern medicine could make a big difference to the lives of the people in such an impoverished area. The kind of development that is conceived of here could be argued to privilege mysticism over Western science, reversing traditional conceptions of development, to the detriment of the community.

SNV cites a number of lessons learned from its experience of CBT, including the obvious one that '[a] Community Tourism Project has more chance of success when based upon skills and attractions that are part of the traditional way of life of the project participants' (SNV/Rozenmeijer 2001: 32). There is a clear logic to this. If tradition is in demand, then the preservation, and even revival, of traditions can be claimed to have the potential to offer limited development, as well as to conserve the past – a sort of symbiosis between the traditional and the modern is established.

In the case of the bushmen of Botswana, SNV also cites a lack of alternative or modern skills within the community. However, it points out that 'the "product" that the community is selling pre-exists – the gathering, hunting and dancing skills which are held by most of the community members of working age' (SNV/Rozenmeijer 2001: 28). Such projects may not be able to 'compete with big safari lodges and luxury tour operators, but they can provide an unusual experience for the more adventurous and inquisitive tourist' (ibid.). Hence, traditional knowledge is the key to this form of tourism development – it bases development on what pre-exists, rather than on changes that may threaten to transform the way the society functions, and it celebrates this as sustainable development.

Overall, SNV's approach echoes that of the two conservation organisations already considered. There is, on balance, a greater emphasis on the value of aspects of tradition as earners of revenue, although the conservation orientation that accompanies this is also strongly in evidence.

Tourism Concern

As a campaigning organisation, rather than one focused on conservation, or one directly involved in the dissemination of development funding, Tourism Concern tends to contextualise its stance on traditional culture within a broader critique of modernity. That traditional cultures are under threat from modern development, and that they need to be protected as a result of this, is discussed explicitly in Tourism Concern's literature.

The emphasis in much of Tourism Concern's output on this issue is that indigenous, rural cultures are under threat from the modern, global economy. Tourism Concern's *Community Tourism Guide* is worth quoting at length on this, as it epitomises their critical take on tourism-as-globalisation:

> One day, somewhere deep in the rainforest in South America or Borneo or Central Africa, a few nervous men and women will step into a muddy clearing in the jungle. Cautiously, they will accept the steel machetes or cooking pots being held out by a government sponsored anthropologist, before hurrying back into the safety of the forest.

> The encounter will not be marked by any great fanfare. It will probably not make the news. Yet it will be a significant landmark in human history. The last 'uncontacted' tribe on earth will have been caught in our global web, and an era of exploration, invasion and global integration that began when Columbus first set eyes on the Americas will be over. For the first time, the entire human race will be connected in one giant, all-embracing cultural and trading network.
>
> As this era of human history comes to a close, we are left with a dominant social and economic system that ignores human and environmental costs. A system that destroys communal life because of its demand for a mobile labour force. That creates mental illnesses and stress by sucking people into huge, anonymous cities. That discourages people from growing their own food because doing so doesn't involve selling anything (and therefore doesn't show up as profit in economic statistics). A system that puts a greater value on a pile of dead wood than a living forest.
>
> (Tourism Concern/Mann 2000: 3)

This lengthy quotation exemplifies Tourism Concern's view of cultural change and traditional societies – modernity, in the form of the global world economy, draws all cultures towards a dominant form of social organisation, which is in their view ultimately destructive.

Tourism Concern's defence of traditional cultures in the face of modern development is a response to the problems that have been associated with development in the past. Modern development has, it argues, ridden roughshod over traditional cultures, hence the need to challenge the way development has been constituted. This very much reflects the neopopulist approach to development considered in Chapter 2 – that the grand schemas of modernisation disregard local culture and traditions, and that the latter should become the basis for development. It is striking that Tourism Concern also regards community tourism as providing an opportunity for the developed world too – the traditional, rural village societies, it argues, have much to teach the developed world:

> In many cases, the indigenous people who live in these [ecotourism destinations] have simply managed their environments better than we in the west, using them productively yet preserving their natural beauty and richness. If we in the west are to relearn a less destructive way of life, then we must learn from their example. Community tours to indigenous communities can be a source of inspiration. They can show western visitors that a 'sustainable lifestyle' and 'living with nature' are practical realities, not just utopian concepts.
>
> (Tourism Concern/Mann 2000: 23)

Hence, here traditional agricultural production is advocated not simply as a valid aspect of a different culture, but as *superior* to methods in the developed world utilising modern technology. This strongly suggests that the defence of tradition is part of a wider critique of modernity.

Notably, tourism itself, in its mass form, is cited as very much part of this destructive process. Tourism Concern argues that cultural traditions are under threat from mass tourism. These traditions can be reduced to 'meaningless tourist attractions' (Tourism Concern/Mann 2000: 13). For example, *The Community Tourism Guide* (ibid.) utilises a hypothetical example of a young boy who can speak some English, and is thus able to earn more money from tourists than established members of the community. Mann poses the following question: 'Will he – and the rest of his village – still defer to the traditional chief or elders, who may have little economic power in this new world?' (Tourism Concern/Mann 2000: 13). Here, the aspiration of the young boy for wealth is cited as creating tension within the existing structures of authority that exist in this particular community. This type of impact is often referred to as the 'demonstration effect' in literature on tourism's impacts (e.g. Nash 1996), and is often seen, as it is by Tourism Concern, as damaging to host societies. However, one could equally argue that the gravitation of younger members of a rural society away from tradition and towards the culture of visitors simply reflects an aspiration to break out of the constraints of tradition and identify with greater affluence. That this affluence generally remains out of reach in impoverished, marginalised societies could be regarded as the root problem, a problem that the limited development characteristic of ecotourism projects does not address.

However, all types of tourism, not just mass tourism, are held to pose a potential threat to traditional societies:

> Even as well intentioned tourists we may still bring our western values and hang-ups: the value we place on money and material gain; our cynicism; our secular scepticism; our belief in individual freedom over communal obligation; our liberal attitudes towards sex and drugs and so on. And tourists with their gadgets and self-confidence can be unwitting propagandists for a western lifestyle.
>
> (Tourism Concern/Mann 2000: 30)

In place of mass tourism, and as a benign alternative to industries such as logging and mining, Tourism Concern advocates community tourism – ecotourism with a strong 'community' orientation. This can, it argues, 'help rural and indigenous communities *preserve their culture*' (Tourism Concern/Mann 2000; my italics). The revenues from community tourism can provide a basis for this – these revenues are dependent on the preservation of tradition as an asset, attracting tourism revenue.

Tourism Concern points out that the interest in tradition from ecotourists creates a basis for societies to benefit economically from their traditions. It also places an emphasis on local knowledge as an asset. It points out that 'tours run by indigenous communities often make a feature of their traditional knowledge of local wildlife or medicinal plant uses, knowledge built up over thousands of years of living in that particular environment' (Tourism Concern/Mann 2000). Ecotourism draws on products that are produced through the application of traditional knowledge, and, as such, can ensure that a greater proportion of the money tourists spend remains with the community – leakages are minimised (Tourism Concern/Mann 2000: 26).

There is a sense in which community tourism attempts explicitly to encourage communities to take on board the preservation of their own traditions. Tourism Concern lists ten principles for its community tourism. These include that '[t]ourism should *support traditional cultures* by showing respect for indigenous knowledge' and that '[t]ourism can *encourage people to value their own cultural heritage*' (Tourism Concern/Mann 2000: 25; my italics).

Elsewhere, in similar vein, it is argued that:

> [m]any indigenous communities have been subjected to years of propaganda from governments, educators and missionaries, telling them that their traditional culture (animism, hunting etc) is primitive, inferior and even evil. Meeting tourists who are even interested in, and respectful of, their culture can be a surprise to many indigenous people. It can *encourage them to re-evaluate their own attitude towards their traditions.*
> (Tourism Concern/Mann 2000: 28; my italics)

A further example of community tourism's role in the preservation of traditional ways of life is given in a case study of a tour in the Siecoya community in the Ecuadorian Amazon, run by Oirana Tours.

> [T]he ultimate aim of these tours is to help the Siecoya maintain their way of life and remain in the forest – not only by generating income, but also by *encouraging the children to value their culture*, by seeing outsiders eager to learn about it, too.
> (Tourism Concern 2000: 34; my italics)

The account also criticises western groups for undermining shamanism, this being part of the traditions of the community.

Tourism Concern is critical of what it suspects to be the 'nature first' outlook of conservation organisations (Barnett 2000). As with SNV, they tend to highlight the possibility for material benefits to be gained through ecotourism on the basis of the community's traditions, rather than the conservation benefits stressed by CI and WWF. Given the more anthropocentric *raison d'être* of Tourism Concern, this is to be expected.

However, there is also, as with SNV, an association drawn between the protection of tradition and the sustainable use of resources. Traditional ways of life are explicitly deemed to be sustainable with regard to the environment, compared with modern development. Indeed, there is an overarching critique of modern development and modern society at the heart of Tourism Concern's outlook – and the defence of tradition is very much a part of this.

IYE

Certainly, the IYE emphasises the importance of tradition in its discussion of cultural facets of development. The *Quebec Declaration* asserts that '[e]cotourism should contribute to [. . .] the cultural integrity of host communities' (UNEP/WTO 2002b: 66), and argues that this is central to sustainable development. The relationship between ecotourism and cultural integrity is stressed throughout the document, and throughout the lengthy *Final Report* (UNEP/WTO 2002a), which identifies this integrity as residing in tradition.

The *Quebec Declaration* argues that ecotourism should:

> [r]ecognise the cultural diversity associated with many natural areas, particularly because of the historical presence of local and indigenous communities, of which some have maintained their traditional knowledge, uses and practices, many of which have proven to be sustainable over the centuries.
>
> (UNEP/WTO 2002b: 66)

Here, indigenous knowledge is associated with being 'sustainable', and hence with all the positive connotations implied by this. This advocacy of traditional knowledge can be understood in the context of the broader critique of modern development mentioned earlier, as imposing Western values and exhibiting a disregard for cultural difference (e.g. McMichael 2000). In contrast to this, the IYE adopts neopopulism, fêting traditional ways of doing things, and argues that, as this comes from the communities themselves, it is more ethical and more democratic.

Yet elsewhere the documentation goes far beyond emphasising local traditions in the spirit of 'recognis[ing] cultural diversity' (UNEP/WTO 2002b: 66), towards advocating it as *necessarily* a good thing. The *Quebec Declaration* lauds ecotourism as being able to 'strengthen, nurture and encourage the community's ability to maintain and use traditional skill, particularly home based, arts and crafts, agricultural produce, traditional housing and landscaping in a sustainable manner' (UNEP/WTO 2002b: 73). This begs the question as to whether the community would freely – outside of the participation exercises conducted by conservation NGOs where funding is invariably tied to predetermined priorities (Ch. 5) – choose 'traditional skill' over technology, 'traditional agriculture' over high-yielding genetically modified seeds, and 'traditional housing' over modern methods better able to stand up to the ravages of nature. Of course, maintaining traditional skills may be attractive to tourists and hence good for trade in ecotourism. However, the IYE presents this as 'sustainable' and as a desirable thing per se, an argument that is itself hard to sustain when we consider that tradition can exercise a conservative grip on how we conceive of social change. As cultural commentator Robert Hewison puts it, discussing the role of tradition and heritage in the British context, if we are 'hypnotised by images of the past, we risk losing all capacity for creative change' (Hewison 1987: 8). This sentiment seems apposite for the emphasis on tradition in the IYE.

Also in the *Quebec Declaration*, it is argued that governments should 'guarantee [. . .] the protection of [. . .] local and indigenous cultures and especially traditional knowledge . . .' (UNEP/WTO 2002b: 67). To argue that sensitivity to indigenous knowledge should characterise development is uncontentious, but a *guarantee to protect it* is quite different. It implies that the decision to protect it is prior to the wishes of the community themselves. Is this simply because of the role of traditional knowledge as a resource attractive to ecotourists, or it is perhaps coloured by disillusionment with modern technology among ecotourism's Western advocates? (Butcher 2003a).

There is also a contradiction in the fêting of indigenous knowledge, and the advocacy of participation, in the IYE literature. The *Final Report* comments that, among delegates, 'it was felt that many local communities did not understand the value of biodiversity very well, and that education was needed' (UNEP/WTO 2002a: 29). Elsewhere, it argues that the IYE should 'promote and develop educational programmes addressed to children and young people to enhance awareness about nature conservation and sustainable use, local and indigenous cultures, and their relationship with ecotourism' (UNEP/WTO 2002b: 69). Given that sustainable development is a hotly contested topic, one wonders what sort of lessons in conservation might be offered to economically poor, rural peoples?

Summary

The case studies exhibit some small differences in emphasis. Notably, the more anthropocentric case study organisations (i.e. SNV and Tourism Concern) place a greater emphasis on the economic development benefits that can be derived from tradition, whereas the two conservation organisations draw a strong relationship between traditional ways of life and environmental conservation. However, both justifications are featured strongly across the case studies, including in the International Year of Ecotourism documentation. That development based on traditional knowledge and conservation goals are mutually reinforcing – a situation strongly associated with sustainable development – is a common feature, and dominates the discourse.

It is also striking that there is little sense of development based on tradition being a stepping stone to the possibility of wider development. Rather, traditional knowledge is viewed as commensurate with sustainable development, the latter viewed as a localised steady state, a harmony between people and environment.

This emphasis on the role of tradition is premised upon a particular reading of culture. It is argued here that culture is viewed as being entrenched in the past, regarded through the prism of cultural relativism, and portrayed as profoundly functional. Further, it is suggested that the emphasis placed on the traditional culture of the communities themselves masks an interventionist approach by the NGOs, whose projects effectively mould tradition around contemporary, externally derived, priorities.

Culture entrenched in the past

It is commonplace for tradition to be regarded as conservative and backward looking – a bastion against progress – when applied to Western societies. Yet in the advocacy of ecotourism in the developing world, tradition seems to be regarded as the favoured basis for development. What the stressing of traditional knowledge represents is an emphasis on *culture entrenched in the past*, rather than culture as the making of the future, a perspective that has clear implications for how one might conceive of development (Butcher 2003a: Chs. 5 and 7).

Expressions of traditional culture evoke a simpler way of life from the distant past, a life from which many in the West feel modern societies have much to learn (Fennell 2003; Acott *et al.* 1998). This anti-modern impulse in the developed world is in part the basis for the growing interest in ecotourism. In Poon's view, tradition is fascinating to the tourist, either simply curious or seeking respite from the modern world, and is considered a central motivating factor for 'new tourists' (Poon 1993). Put simply, the association of modern Western societies with a lack of authenticity leads to a search for authenticity in the past. 'Community', 'spirituality', 'harmony' and 'closeness to nature' are prominent in this search, and a romanticised version of rural developing world communities provides these in abundance (Butcher 2003a: Ch. 5).

Of course, there can be little objection to such a reading of culture by tourists. Indeed, holidays have always been associated with romanticism, experimentation with different ways of living and imagination (Feifer 1985). However, in this instance, the emphasis on traditional aspects of culture becomes a constraint on the development on offer for rural developing world communities. Development strategies based around such a reading of culture are never likely to propose much beyond a limited development based upon what communities already have by way of 'cultural resources', which in this case are rooted in tradition – indeed, such strategies eschew change beyond this. The consistent invocation of tradition as central to the culture of the communities suggests communities as rooted in a history that leads nowhere, that cannot develop *beyond* tradition. The linking of projects to distinctive and severe natural limits, the lack of reference to development beyond the limits of the project, and the labelling of this scenario as sustainable development, strongly suggests that traditional knowledge is being invoked in the case studies to sustain the way things are, and to eschew substantial change.

The influence of cultural relativism

There are plenty of allusions to cultural relativism in the NGO case studies, and in the advocacy of ecotourism generally. WWF's stance on biocultural diversity, referred to earlier, is a clear expression of this – it views culture as intimately linked to the specific relationship of a community with the environment (Maffi and Oviedo 2000). Culture is thus regarded as diverse in the same way as biology is diverse – there is little room in this view for aspirations for human equality or to make scientific advances from the more developed parts of the world universally available. Instead, our culture is defined by our differences, not by what we might have in common.

Elsewhere in the case studies, cultural relativism is strongly implicit or explicit. The view from CI is that ecotourism is worthy in that it supports the 'current culture and heritage of people and the diversity of nature at a particular destination' through placing values on specific 'cultural attributes' (CI undated e), while SNV contends that ecotourism in Tanzania should be a 'source of cultural preservation for the Bushmen' and 'an income generating project that is based on knowledge that people already possess' (SNV/Rozenmeijer 2001: 26). Invoking cultural sensitivity, Tourism Concern asserts that the 'ultimate aim' of community tourism in Ecuador should be 'to help the Siecoya maintain their way of life and remain in the forest'. The IYE seems to concur that cultural diversity is in and of itself a good thing, regardless of any desire the Siecoya or any other indigenous group may have to 'come out of the forest'.

A clear theme in all the case studies is development based upon the specific way of life and traditions of the local community, not as a first step towards opening opportunities for higher levels of economic development (which would destroy traditions) but as the end, sustainable development, itself. As such, the advocacy of traditional knowledge as central to development through ecotourism draws heavily on cultural relativism. Raymond Williams has argued that cultural relativism is characteristic of a distinctly anthropological conception of culture (Williams cited in Young 1995: 44), and Milton concurs, pointing out that anthropology has been influential in thinking on rural development since the 1980s, and has brought with it cultural relativism as an influential point of reference (Milton 1996).

Indeed, anthropology has provided many of the tools and terms of reference utilised in the study of rural development (Milton 1996), and specifically in the debates on the impacts of tourism development (Butcher 2003a; Burns 1999; Smith 1989; Graburn 1988). Discussions of tourism's impacts on the developing world draw directly on ideas such as the demonstration effect, acculturation, commoditisation, staged authenticity and cultural imperialism, all terms either anthropological in origin or widely written about in relation to tourism by anthropologists (notable examples include Nash 1996; Selwyn 1996).

The anthropological outlook has merit as it may lead to greater sensitivity to different ways of life and how they may be affected by economic and social change (Cole 1997). Yet at the same time its cultural relativism can be a barrier to arguing for greater material development and greater equality between the developed and developing worlds. For cultural relativists cultures are more than different ways of life – they become fundamentally *different ways of knowing* that can only be properly understood 'in their own terms' (Milton 1996: 19). The logic that flows from this is that cross-cultural comparison is impossible and the problems of cross-cultural communication severe. This inevitably leads to a reluctance to argue for development based on modern technology. Indeed, some have questioned the very idea of development, seeing it as a totalising discourse that cannot deal with the fundamental importance of cultural difference (see Potter *et al.* 1999: 9–14 on anti-development).

Put simply, anthropologists' taking of the side of the other culture, laudable in itself, is accompanied by a tendency to see a society as defined by its differences from other societies. In relation to rural, traditional societies, anthropology's influence has sensitised us to problems arising from inter-cultural communication, but can deprioritise access to the fruits of modern science for all, the latter viewed as a cultural imposition, or as not constituting 'appropriate' technology. There is certainly no sense at all in which the invocation of traditional knowledge in the advocacy of ecotourism is viewed as a stepping stone to greater development possibilities involving the utilisation of modern technology – it is taken to be, either explicitly or implicitly, appropriate development.

This cultural relativism often takes the form of counterposing indigenous knowledge systems to alien impositions from richer nations, and this is a feature of the discourse outlined earlier in the chapter. Development theorist Escobar articulates a view that underpins many of the pronouncements about traditional knowledge in the advocacy of ecotourism: 'Development has relied exclusively on one knowledge system, namely the modern Western one. The dominance of this knowledge system has dictated the marginalisation and disqualification of non-western knowledge systems' (Escobar 1995: 13). Escobar views the problem of development in terms of different 'knowledge systems', Western and non-Western, effectively disaggregating knowledge on the basis of culturally defined groups.

The alternative proposed by Escobar – which chimes with the advocacy of ecotourism – is nothing less that the 'remaking of development' which must 'start by examining local constructions, to the extent that they are the life and history of the people, that is, the conditions for and of change' (Escobar 1995: 98). In this formulation non-Western societies should follow their own specific trajectory based upon their own local traditions, and eschew the 'knowledge systems' imposed by the West. In effect, he relativises scientific knowledge, taking modern science and traditional knowledge as different systems, rather than one being an advance of the other, better able to explain and act upon the world, and having the potential to bring benefits to healthcare, food production etc.

Escobar is wrong to argue that traditional knowledge is a different system of knowledge from modern science in the sense that both arise from the same impulse to understand and organise the natural world around human needs and wants, be it under very different circumstances. Yet modern science involves a quantitative and also qualitative advance on traditional knowledge: quantitative in that it has developed through and out of traditional knowledge, qualitative in that it enables modern societies that can utilise it to plan their relationship to the natural world in a broader, systematic way that local, traditional societies cannot conceive of.

Proponents of traditional knowledge argue that its importance is that it is deeply embedded in context, in the 'local world' – it may not apply elsewhere. In making the case for traditional knowledge, Pretty (1994) and Agrawal (1995) distinguish between traditional knowledge and Western scientific knowledge on the basis that the former is intertwined with the lives of the community, whereas the latter is separate from it. Yet this feature marks out the limitation of traditional knowledge vis-à-vis modern science. Modern science has developed the capacity to systematise and communicate new knowledge (from whatever source) – its frame of reference is global. It has the capacity to be able to take the pre-existing position of a society as the starting point, and propose opportunities to improve yields, to reduce the physical burden of agricultural work, to promote mobility and improve resistance to disease. As such, modern science reflects the highest form of knowledge arising in the course of human development as a whole.

Hence, traditional knowledge is a limited and stunted level of knowledge, not an alternative 'system of knowledge' to modern science. The context of traditional knowledge is poverty. Logically, if the context within which knowledge is generated is to change – if poverty is to be ended through development – we cannot expect traditional knowledge to survive, nor should we necessarily bemoan its loss. Ecotourism's advocates insist it should serve as the basis for sustainable development.

Societies are divided by their access, or lack of it, to the fruits of human development embodied in modern science and technology. The denial of these benefits to people in the developing world is a far bigger problem than the unwanted imposition of alien 'knowledge systems'. The discussion of traditional or indigenous knowledge systems turns reality on its head in suggesting that the key problem is too much emphasis on modern science and too little on traditional knowledge. Stating the case for a humanist perspective, one that is optimistic about the capacity of science and technology to transform societies for the better, goes against the grain of much debate on rural development via ecotourism. The influence of cultural relativism has made it difficult to argue the simple humanist aspiration that the benefits enjoyed in one society should be made available to others. It is hard to sustain the deference to tradition when the fruits of science and technology are so little in evidence in many parts of the developing world . . . and where health, literacy and life expectancy are lower as a result.

And yet ecotourism takes traditional knowledge as the basis for development – not as a starting point, but as its basis per se. This is clear in the discourse set out earlier. Even where there is the hint of a possibility of wider development, this is implicitly very limited by the emphasis on a sustainable localised relationship between the community and the environment. This is not to argue that a sensitivity to existing local knowledge is not important. However, the cultural relativism of ecotourism's advocates seems to defer to culturally specific knowledge embodied in local tradition, not only as a point of departure, but as a principle. This emphasis on tradition (the way things have been in the past*)* constrains any discussion about substantial change (the way things could be in the future), which involves envisioning development based on access to the most modern technology and scientific knowledge.

Culture as functional

The emphasis on tradition presents us with a notion of culture that is *functional* with regard to the operation of the community *as it is constituted*. This functional character of culture is central to the advocacy of ecotourism in the case studies and is in many ways straightforward and uncontentious. It is a truism to say that aspects of culture help to cohere societies, and socialise new generations into the values, norms and social roles characteristic of the society.

However, if we are studying societies in the context of *social change*, or *development*, the functional conception of culture arising from the emphasis on tradition may carry profoundly conservative assumptions. This point can be illustrated with reference to the biological analogy often invoked to explain functionalism in the social sciences. Culture is often seen as analogous to the human body, with different organs (cultural norms, perhaps) enabling the overall functioning and survival of the body (the society) (Rojek 1995: Ch. 3). Change to aspects of culture wrought by development is seen as upsetting the functioning of the society more generally.

Yet the analogy is also indicative of the limitations of functionalism. The human body may function biologically, but human beings, and their societies, function *socially*. Likening human culture to biology inclines towards a *naturalisation* of culture, suggesting culture should not change substantially, and only evolve slowly and gradually. Indeed, in the case studies, tradition is often presented as a natural order of things, an order under threat from the modern world, and hence in need of support. Part of the rationale for ecotourism projects is that they offer such support.

This is clear in the case study summaries. Tourism Concern's pronouncements on encouraging tradition to discourage the adoption of Western lifestyles, SNV's view that the emphasis on tradition can preserve the status of traditional leaders and ensure sustainable development, and the unequivocal opinion expressed in the *Quebec Declaration* that ecotourism can support tradition, which in turn supports sustainable development, are examples of a pervasive view.

Indeed, the whole philosophy of ecotourism – that it can create a symbiosis between human development and biodiversity conservation – suggests a view of culture as functional, its role being to sustain a particular relationship with the local environment. This view is common to the many academic advocates of ecotourism (Fennell 2003; Goodwin 2000; McLaren 1998). Development is constructed as a virtuous circle in which ecotourism supports traditional culture, which in turn supports conservation, which in turn sustains traditional culture and so on. Yet the circle reflects a static view of society – substantial change that would involve a transformation of the relationship between people and environment is off the agenda. Essentially what is being proposed is a localised steady state between people and environment. In this sense, to see culture as functional in relation to societies *as they are* is to restrict a discussion of societies *as they could be*. The conception of human agency at the level of the community is thus diminished – while tradition can play the role of cohering a society and shielding it from malign influences, it can also be a bastion against aspiration and progressive change.

The term 'agency' is used widely in sociology with slightly different emphases. Most commonly, and here, it is utilised with regard to the 'structure–agency' issue that is prominent in social and political theory; 'structure' emphasising the determining contextual features of a society, features that determine or constrain human action, and 'agency' emphasising undetermined human action (Giddens 1991). Put simply, functionalism relies on a conception of culture as structure, whereas agency emphasises the ability of individuals and groups to act upon their society, their context, and to change it (ibid.). Adopting functionalism, culture can appear to exist outside of, and prior to, the inhabitants of the local community – it is an external force, holding the society and the people within it together. The logic of functionalism is that culture makes man, rather than man makes culture.

An irony here is that agency, through the emphasis on local tradition (as well as community participation), is constantly invoked in the case studies and by neopopulist advocates of ecotourism. It is *the community's* culture, *their* tradition and hence *they*, the host society, who make their own future. Hence, much of the rhetoric suggests an enhanced ability of the community to make and remake their own society. Yet this progressive rhetoric masks a particular view of culture, one that ties it to the pre-existing way of life, to the pre-existing relationship to the local environment, to traditional knowledge – it is a profoundly functional conception of culture, suggestive of a steady state between people and the local environment.

Against stasis

The argument that an emphasis on tradition can embody a static view of society that denies human agency, as set out above, has been made elsewhere (notably, Hewison 1987). A similar argument sometimes put against ecotourism is that it treats the host society as a museum piece, to be gazed upon as an example of ancient societies, of the past, while the community's ability to adapt to and benefit from change is played down (Butcher 2003a; Mowforth and Munt 1998).

However, the dual emphasis on conservation *and development* in the case studies would suggest that the NGOs concerned are, in fact, proposing something more progressive than this. This is a sentiment expressed occasionally by some of the case studies, mainly by those of a more anthropocentric outlook, SNV and Tourism Concern. These organisations are more concerned with well-being – this is central to their aims. For example, Tourism Concern constantly emphasises the role of 'people' and 'participation'. Also, SNV argues with regard to its work in Botswana that it is not the aim 'to keep Bushmen in stasis – the skills and money they gain can be used in any venture they might choose' (SNV/Rozenmeijer 2001: 27). It is hoped that the Bushmen will be able to 'pursue self development' and, in that sense, community-based tourism may be a 'jumping off point' (ibid.).

Within the case studies, however, there is little if any sense of ecotourism being the beginning of possibilities for more thoroughgoing development – ecotourism ICDPs are not discussed as a stopgap until aid budgets are more substantial, or as a stepping stone towards developed world status. Indeed, such a discussion is conspicuous by its absence. This is both surprising but also consistent with the conceptual underpinnings of ecotourism as a development tool, of which an emphasis on traditional knowledge is a key aspect.

And, if ecotourism can be seen as a 'jumping off point', where can the community jump to? Traditional knowledge is firmly rooted in subsistence agriculture and craft production, which itself reflects the direct relationship of the community to their environs (unmediated by modern technology or a wider division of labour). If development deemed sustainable is based on tradition and conserving natural resources, or natural capital, then there is little basis for going beyond this in pursuit of wider development. Wider development would require the transformation of natural capital, and the transformation of the relationship between the community and its natural environment. The philosophy behind ecotourism seems to preclude this, premised as it is on reducing the 'impact' on the host society (Fennell 2003). Hence, a position that appears anthropocentric, in that it suggests people can take charge of their own development (thus suggesting development possibilities as being more open ended), may in practice foreclose a discussion of development alternatives beyond those corresponding broadly to the status quo.

Some looking at the issue clearly recognise the limitations in economic development on the basis of tradition, and propose a more development-oriented ecotourism. This is most clearly evident in the developing discussion about 'pro-poor tourism', an initiative pioneered in the UK by the Department for International Development in conjunction with the think tank the Overseas Development Institute (see Ashley *et al.* 2000).

Pro-poor tourism takes issue with what it rightly sees as 'a defensive or protectionist approach: "preserving local culture", "minimising costs"' in the language of sustainable tourism (DfID 1999a). In contrast, pro-poor tourism is presented as being about the expansion of opportunities (ibid.). It is also presented as a broader approach than community-based tourism, as it prioritises the links between impoverished communities and the formal sector (ibid.).

However, this more anthropocentric presentation of ecotourism may be less differentiated from those evident in this study than at first apparent. One of the justifications given for utilising tourism in this way is that '[t]ourism products can be built on natural resources and culture, which are often the only significant assets the poor have' (DfID 1999a: 2). Yet this lack of resources defines their underdeveloped status. The pro-poor tourism approach works around this to engender limited development in rural areas.

Indeed, one departmental workshop paper on the subject, entitled *Sustainable Tourism and Poverty Elimination*, begins with a quotation from WWF, suggesting that localised natural factors constrain the extent to which poverty can be tackled: 'Sustainable Tourism is tourism and associated infrastructures that, both now and in the future, operate within *natural* capacities for the regeneration and future productivity of natural resources' (cited in DfID/DETR 1998; my italics). Yet in what sense are there natural limits in the fashion implied? The limits to development in the developing world are better regarded as *social* in nature rather than rooted in natural processes. They are a product of unequal economic and political relationships and, more immediately, the burden of debt and the dearth of inward investment and of aid itself. Few in the more developed countries live within limits defined by their specific relationship to their immediate environment in the way DfID's advisers seem to be advocating here in the name of sustainability.

An alternative formulation of the argument that ecotourism is not anti-change is that it enables tradition to remain viable in the modern world, and hence enables communities to retain what they value from the past as society changes. This view is articulated by Dean MacCannell, one of the foremost theorists of tourism as a human activity (MacCannell 1992). Referring to the impact of tourism on the Masai peoples in Kenya and Tanzania, MacCannell argues that, through tourism:

> the assimilation of primitive elements into the modern world would allow primitives to adapt and coexist and earn a living just by 'being themselves', permitting them to avoid the kind of work in factories and as agricultural labourers that changes their lives forever.
>
> (MacCannell 1992: 19)

For MacCannell, Masai traditional culture can be *sustained* in a form that does not preclude development by a degree of commercialisation arising from tourism.

Yet, although such a conception clearly allows for the possibility of progressive cultural change and development, it retains a deep-seated antipathy to modern societies, with their factories and commercial farms. Following MacCannell's logic, Masai tribesmen who aspire to integrate into mainstream Tanzanian society are no longer 'themselves', having lost their traditions. Yet the whole history of development has been characterised by movements from the country to the city, by assimilation of different cultures, the formation of social classes and industrialisation. This legacy is viewed with great scepticism by some advocates of ecotourism as a development tool. Indeed, the advocacy of ecotourism reflects a deep-seated disillusionment with modern, Western societies, perhaps best exemplified by the summary featuring Tourism Concern on pp. 109–12.

Cultural intervention

According to one authoritative definition, 'tradition' refers to '[a] set of social practices which seek to celebrate and inculcate certain behavioural norms and values, implying continuity with a real or imagined past' (Marshall 1998). This is apposite – the support for tradition is with a view to inculcating 'sustainable' behaviour – behaviour that is in keeping with the prior goals of the projects. It is not, as it is often presented, a championing of the unfettered agency of the community. Indeed, as the case studies suggest, ecotourism ICDPs involve an intervention into culture, to mould and adapt tradition to fit the project's aim.

In the case study summaries for WWF and CI, traditions are supported as they provide a bastion against developments deemed to be damaging to important ecosystems. When aspects of traditional culture may threaten conservation, then the projects include measures to *modify* tradition. A further example of this is the Campfire scheme in Zimbabwe – a scheme in which WWF has been centrally involved, and one also featured favourably in its literature by Tourism Concern (Tourism Concern/Mann 2000) – whereby communities are encouraged not to hunt by the prospect of ecotourism revenues, revenues that may yield more wealth and food than the hunted animals' carcasses. Other examples of this tendency are featured in the case study summaries.

Ecotourism ICDPs constitute an intervention in the cultural lives of the communities, in just the same way as other developments do – developments of which those advocating ecotourism are often highly critical. The act of supporting a tradition, through enabling it to act as an asset for the community, inevitably changes the character of that tradition. This straightforward point is worth making here if only because the advocacy of ecotourism sometimes refers to the view that the emphasis on preserving traditions protects communities from external cultural influences. Indeed, all the summaries presented in this chapter allude to this sentiment. Yet the conscious act of preservation, sponsored through aid funds, is itself in an important sense an external cultural influence. Again here, the association of tradition with the agency of the community itself may be questioned.

Moreover, this is an intervention that assumes the community has fixed income needs – that the small development benefits on offer will satisfy their aspirations, and they will moderate hunting and organise their community to promote conservation once a certain level of basic needs has been met. Evidence from ICDPs suggests that, in fact, communities aspire to develop their wealth further, beyond the small benefits on offer from the projects (Mogelgaard 2003). In economic terms, there is an income effect – small gains in income from ecotourism revenues encourage people to spend these gains on improving their lives further, which can encourage hunting to continue to meet the increased demand. This obviously compromises the conservation aim of the project. The logic of ICDPs is that such desires for continuous development have to be reined in, in the interests of a harmonious relationship to the environment (ibid.). The emphasis on local tradition plays a part in portraying this as a 'people-centred' approach.

The environmentalism of the poor?

Some writing on rural development argues that while industrial processes have an exploitative, commercial relationship with nature and hence work against it, some non-industrial societies have a spiritual tie and live in harmony with nature (Milton 1996; Ellen 1986). In similar vein, some see non-industrial societies as models for 'sustainable' societies (Paehlke 1989: 137–41). Specifically, this point has been made in relation to ecotourism – tourists can learn from the harmonious relationship the communities visited have with their natural environment (Fennell 2003; UNEP/WTO 2002a; Tourism Concern/Mann 2000; Acott *et al.* 1998). Indeed, this is fairly explicit throughout the advocacy of ecotourism.

Yet in terms of formal belief, there are non-industrial societies that do not recognise a human responsibility towards the environment at all (Milton 1996: 133). Forms of conspicuous consumption are in evidence in many pastoral societies – there is no conservation ethic in the developing world that we can counterpose to a consumer-oriented developing world (Milton 1996: 139). Indeed, Milton comments that: 'It might come as quite a shock to western environmentalists to learn that some of the least environmentally damaging societies are culturally closer to industrial entrepreneurs, in some ways, than to themselves!' A low impact on environment can exist alongside a culture entirely open to the benefits of high impacts (Milton 1996: 135), and some cultures even see nature as infinitely generous (an idea that finds little support in the developed world!) (Milton 1996). Of course, a direct comparison with the developed world would be fatuous, other than to point out that the existence of a set of environmentally benign ideas about how to live, running counter to an environmentally destructive 'culture of industrialism' (Milton 1996: 140), does not exist in a rural developing world or anywhere else – there is no 'environmentalism of the poor'.

Milton proceeds to ask why the myth of what she terms Primitive Ecological Wisdom, continues to exert influence, and lists among her views the possibility that environmentalists fail to distinguish between culture as in what people 'think, feel and know', and culture as simply 'things people do' (Milton 1996). She argues that the 'things people do'– embodied in traditions, traditional knowledge, traditional agricultural methods etc. – become a limiting point of reference in the culture discussion, as it is taken to reflect the deeply held beliefs and desires of the community, or what they 'think, feel and know'.

Yet that 'the things people do' may indeed be benign towards the environment is less the outcome of the agency of local people expressed through what they 'think, feel and know', and more straightforwardly a product of poverty itself. Milton argues that the reasons non-industrial societies appear benign towards nature are rather prosaic, and include there being few people partly due to disease and short lives; the relative isolation of communities, with a lack of opportunities for trade; and a simple lack of technology which may preclude damaging nature (Milton 1996: 122; Ellen 1986). Hence, any ecological balance (or symbiosis between local culture and environment) may be incidental rather than a goal actively pursued (Milton 1996: 113).

Milton's analysis gives the lie to the 'environmentalism of the poor', the view that economically poorer societies embody traditions and belief systems that have much to teach the developed world about the value of natural resources and how to live a sustainable life. And yet this latter view is a central part of the advocacy of ecotourism. Indeed, ecotourism is not only deemed to have an important educative function for Western tourists and their societies (Fennell 2003; Tourism Concern/Mann 2000; Acott *et al.* 1998), but is held up in this respect as an example to the rest of the tourism industry as exemplary sustainable development (perhaps most notably in the IYE documentation [UNEP/WTO 2002a]).

Another probable reason why the myths about traditional societies persist is that developing world societies have allowed themselves to be depicted in this way to attract the support of wealthy environmental organisations (which in the case of ICDPs play a significant role in spending rural development funds). If wealthy funders are positively attuned to a discourse about traditional knowledge, then it could be that communities in the developing world will adopt this rhetoric too, in order to gain access to aid funds – this is a process well elucidated in a number of contributions to *Civil Society: Challenging Western Models* (Hann and Dunn (eds) 1996) in relation to civil society discourse in development aid. In similar vein, it is likely that communities offered funding on the basis of prioritising traditional knowledge will buy into this if this is all that is on offer.

The emphasis on traditional knowledge – made in the West?

Opposition to modernity has been a key theme in conservation thinking in the West (Pepper 1996). 'Tradition' has been seen as an oppositional category to 'modernity', and is invoked to support the notion that conservation has a natural affinity with indigenous people and rural dwellers in the developing world (Neumann 1997). Yet, as alluded to already, this may be based on a mixture of myths and debatable presuppositions about the developing world. Milton's view, alluded to earlier, is that there is a sense in which it is a view constructed in the West and superimposed upon the South. This is a wider version of an argument central to this study – that a discourse that invokes the culture of the developing world community has relatively little to do with the needs and wants of that community, other than those that can be expressed through staged and limited 'participation' exercises.

John Reader, in his *Africa: a Biography of a Continent* (Penguin 1998), argues that it was Africa's colonisers who put tradition at the centre of the Western vision of the continent thus:

> The colonisers claimed that they were merely confirming the significance of existing traditions, but traditions in Africa (and everywhere else for that matter) are merely accepted modes of behaviour that currently function to benefit society as a whole. They persist so long as their benefit is evident, and fade away when it is not. No tradition lasts for ever. Change and adaptability are the very essence of human existence – nowhere more so than in Africa. The paradox is painfully evident: by creating an image of Africa steeped in unchanging tradition, the colonisers condemned the continent to live in an unreconstructed moment of its past, complete with natives in traditional dress, wild animals and pristine landscapes.

Today, tradition is commonly invoked as central to rural development through ecotourism in a not dissimilar fashion . . . and with not dissimilar dangers. Just as African tribalism was consciously crafted by the colonialists, rather than being deeply rooted in Africa's past (Reader 1998: Ch. 51), so too is today's emphasis on indigenous traditions very much a product of modern thinking on green, 'sustainable' development in the West. The debates look very different – colonialists asserting a sense of superiority over 'tribal' peoples, and modern green development advocates defending African traditions from the assault of modernity would, on the face of it, seem to be taking different sides. However, in both cases, the result is to fetishise and reify tradition. The shared assumption is that African traditions tie the potential for or desirability of change to a pre-existing harmony with the community's local, natural environment.

With regard to contemporary discourse, Milton (1996: Ch. 4) argues that the myth of Primitive Ecological Wisdom is widespread and influential in the West, and is part of a romantic tradition of idealising the natural, and seeing rural communities as inhabited by the 'noble savage' (ibid.: 109). A variation of this view is developed in Campbell's *Western Primitivism: African Ethnicity* (1997), a study that situates the emphasis on ethnicity in the contemporary discussion of Africa firmly in the culture of the West. As a sense of mission and self-confidence in the West has ebbed, as disillusionment with the fruits of modern society has grown, a desire for a closer relationship to the natural world (albeit one that sits alongside the adoption of certain types of modern technology – those applied to the individual rather than society) has become a feature of Western culture. As Campbell argues, although modern champions of developing world communities would baulk at the suggestion, some articulate a modern version of the 'noble savage' outlook associated with colonialism and imperialism of the nineteenth century (1997: 38). Whereas imperialists had a sense of their cultural superiority over Africans, who were deemed noble, but ultimately savage, today's cultural relativists emphasise the richness and 'sustainability' of rural communities' way of life. The common feature of these views is that both see the community's culture as rooted in its relationship to the land, a relationship mediated through tradition, and see neither the possibility nor desirability of changing this substantially.

Conclusion

The emphasis on local tradition initially appears wholly progressive. First, critics of the modernisation development paradigm have long claimed either that development has ridden roughshod over local cultural values, or that it superimposes a Western culture onto communities who may not have chosen this (e.g. McMichael 2001). Specifically, the importance attributed to traditional knowledge appears to resolve some of the problems associated with development as identified by neopopulist thinkers such as Chambers (1997), Hettne (1995), and, in the field of tourism, Scheyvens (2002). Traditional knowledge invokes the agency of the community – it is *their* knowledge, not a Western imposition. However, agency is restricted to the community's way of life, a way of life shaped by its direct relationship with the natural world. This functional approach cannot envision the community's agency extending beyond a pre-existing 'way of life' into the realm of substantial and social change. Conceivably, the emphasis on tradition may orient the discussion of development away from, for example, the usage of more modern technology, technology that could transform tradition and potentially offer benefits based upon this.

This criticism of the modernisation paradigm is prominent within the advocacy of ecotourism too (e.g. Neale 1998; Croall 1995), and in many books on tourism's impacts on societies and on development (Scheyvens 2002; Honey 1999; McLaren 1998; Cater and Lowman 1994; Krippendorf 1987; Turner and Ash 1975). Ecotourism is presented by its advocates as a counter to a cultural arrogance they associate with modernisation, and as more rooted in the desires of the community, more responsive to the rhythms of their society. The instinct in the advocacy of ecotourism is to defend local traditions in the face of development as modernisation.

The neopopulist critics cite the tendency for modernisation as development to ignore the question of culture. However, the philosophy behind ecotourism, evident through the case studies, goes further than simply arguing for greater cultural awareness. The philosophy of the projects here seems to be the *inversion* of the modernisation paradigm, holding that development has to be informed by, and limited to, that which conforms to prior knowledge and traditions. This is the legacy of cultural relativism.

Many of today's critics of tourism subscribe to the view that it can fictionalise situations involving cultural contact in a way that, as MacCannell argues, 'assum[es] the superiority of the west' (MacCannell 1992: 295). The advocacy of ecotourism is certainly a response to this perception of cultural arrogance on the part of tourism and cultural degradation for host societies (Tourism Concern/Mann 2000; Neale 1998; Krippendorf 1987), and its advocacy adopts a precautionary approach to cultural contact. However, the advocacy of ecotourism creates its own fiction based on the *inferiority* of the West – its advocates elevate the traditional characteristics of poor host societies and decry their own. Although presented as sensitivity to these communities, this approach implicitly restricts discourse on development, and potentially development itself, to what can be achieved on the basis of very little substantial change to rural developing world societies.

The emphasis on traditional culture presents *culture as functional* – traditional knowledge can play a role for the society or the community as it is constituted, and as such can clearly be argued to be positive in this respect. However, parallel to this, this functional view of culture stands in the way of any change that may transform the society's relationship to the natural world and to its own traditions. As such, the emphasis on traditional knowledge may potentially contribute to a particular, and narrow, discussion of what is possible and what is appropriate in the rural developing world. This reinforces the presentism discussed in Chapter 1.

This is evident if we simply consider that historically all developed societies have undergone transformative change in the course of development – be it in the form of an industrial or a technological revolution. All such change is destructive of tradition, but creative in the sense that it opens up new possibilities for people based on higher levels of production and technology. It is striking that it should be in the rural developing world that sustainable development is most closely associated with a form of development that seems to favour tradition over change in this way.

Having considered the broadly 'cultural' aspect of the advocacy of ecotourism, Chapter 6 looks at the 'environmental' assumption that is also a central feature – that the non-consumption of natural capital constitutes exemplary sustainable development in environments deemed fragile.

6 Natural capital in the advocacy of ecotourism

Introduction

It has been established that ecotourism is advocated as having the potential to constitute exemplary sustainable development in the rural developing world. An important part of this advocacy is that other forms of development are deemed to be less sustainable with regard to their impact on the environment. Projects implicitly and explicitly base development upon the non-consumption of natural resources, or natural capital, rather than through the transformation of nature in the course of economic development (Fennell 2003; Boo 1990; Ziffer 1989). Hence, they advocate as sustainable development a type of development in marked contrast to the experience of the developed world.

This chapter begins with a discussion of two key concepts, 'natural capital' and 'environmental fragility', each of which, it is argued, is an important aspect of the advocacy of ecotourism as sustainable development.

Following this, there is a summary of findings from the case studies with regard to their approach to the natural environment. This summary focuses on the way the NGOs extol ecotourism for its *non-use* of natural capital (Fennell 2003; Boo 1990), and note that this is often done by favourable comparison to other forms of development that consume, or 'use up', natural capital to a greater degree.

The summary indicates some important differences and continuities between the case studies, and these are subsequently analysed. In this analysis it is argued that, though different NGOs have differing emphases in terms of how they rationalise environmental fragility, they tend to arrive at a very similar conclusion – that transformation of the environment through development is inherently detrimental. In their advocacy of ecotourism in the rural developing world, they therefore share the view that sustainable development is development through the non-consumption of natural capital. Thus, they adopt what has been termed a 'strong sustainability' approach to sustainable development (Ekins *et al.* 2003: 167; Beckerman 1995 and 1994), an approach in which natural capital is viewed as unable to be compensated for by technology and development. Strong sustainability, it will be argued, is a *particular* version of sustainable development, one that could be regarded as strongly ecocentric, and also as pessimistic with regard to the outcomes of economic development (Beckerman 1995 and 1994).

It is further argued that this assumption – that development should be, by and large, on the basis of non-consumption of natural capital, rather than through its transformation – severely constrains any discussion of development possibilities.

What is natural capital?

The concept of natural capital is strongly invoked, both implicitly and explicitly, in the advocacy of ecotourism as sustainable development in the developing world (Fennell 2003; Honey 1999; Boo 1990; Ziffer 1989). This section looks at this concept in order to gain a clear understanding of it, prior to considering its implications.

Conceptions of capital have generally referred to the creation of value through the transformation of the natural world into means of production and products themselves. This is the case in the classical economic theories of Smith, Ricardo and Marx, and in the subsequent neoclassical variations (Maunder *et al.* 1995; Rubin 1979; Galbraith 1969). Natural capital, on the other hand, refers to biophysical and geophysical processes and the results of these processes – fish in the sea, timber in the forests, oil in the ground – and the relationship of these to human needs over the long term (Tacconi 2000: Chs 3 and 4; Berkes and Folke 1994). For example, one could argue in this vein that cutting down the rainforest should be seen as running down stocks of natural capital, even though from a purely commercial point of view the trees may have no value outside of what they can yield once human capital and capital in the form of machines – labour and sawmills respectively – have been applied to them, and they have been sold on markets.

The natural world does feature in neoclassical economic theory, as the category 'land'. The creation of value comes about through the combination of land, labour and capital (and in some formulations, entrepreneurship). Within this, land attracts rent (there is a market for it). However, this arguably takes no account of any potential human welfare gains through the *non-use* of the land, and the non-disturbance of the natural processes residing in it. These would include, for example, the genetic diversity within an ecosystem and the potential for this to yield up benefits to medicine, or the role of forests as 'carbon sinks' (Pearce and Moran 1994).[1] Any consequent benefits for communities may be regarded as 'capital cheap' – an important consideration in indebted countries.

Post-Second World War developments in economic theory were largely silent on the issue of natural resource conservation up until the 1970s (England 2000: 425). For example, neither the Harrod/Domar Model of a dynamic, uneven relationship between capital investment and growth, nor Solow, who responded that this unevenness was not inevitable, consider it (ibid.).

In the 1970s, the 'limits to growth' school emerged, positing environmental limits to economic growth, and reflecting a wider recognition of environmental concerns (Adams 2001: 46–7; England 2000: 425–31). Opponents of this school accepted that environmental effects of economic growth were an important issue, but generally emphasised the ability of societies, through technological advance, to offset declining resource stocks (England 2000: 425–31).

The idea of natural capital itself was first introduced in the 1980s, reflecting a 'new, more ecologically aware thinking in economics' (Akerman 2003: 431; see also Tacconi 2000: Ch. 4). Previously, welfare economics – a relatively minor field within economics – had considered the environmental effects of economic growth, but effectively treated these effects as *externalities*, or by-products of economic activity (Akerman 2003: 431). The invocation of natural capital, by contrast, was part of a new 'ecological economics' that emerged in the late 1980s, as a distinctive 'interdisciplinary bridge between economics and ecology' (Akerman 2003: 434; see also Tacconi 2000: Ch. 3). This school of thought sought to address the emerging imperative of sustainable development through combining ecological and economic perspectives in theory. The use of the term natural capital marked an attempt to make the natural world integral to economic thought and to national accounting.

The term natural capital has a strong normative edge to it – it is often invoked in the advocacy of how things *should* be. It challenges traditional neoclassical economic thinking, positing nature *in and of itself* as a source of welfare, rather than a relatively passive element in the production process (Akerman 2003).

The importance of natural phenomena and natural processes may be regarded as of a different order from more traditional capital theory, the latter readily understood in terms of monetary exchange value realised through the market. This means that, while natural capital may be seen as playing an important role as *metaphor* (Ekins *et al.* 2003: 169), pointing to the importance of natural processes, it may equally be seen as 'analytically weak' (Akerman 2003: 435). For this reason, it has been argued that natural capital may be best understood as 'a linguistic device, a fluid object' (ibid.: 439), brought into play to push environmental conservation onto the economic development agenda. For example, prominent advocate of the efficacy of natural capital, Robert Constanza, sees it very much in this way, in the context of a critique of neoclassical economics and its limitations (Constanza 2003: 19–28).

The importance of natural capital may be less that it provides a precise guide to action on the environment, and more that it emphasises broad natural limits to the endeavours of human societies to develop economically through transforming the natural environment (Akerman 2003; Tacconi 2000: Ch. 3). Though many would accept that there are such natural limits, what they actually are is contested. Many advocates of natural capital argue that these limits have already been surpassed (e.g. Gowdy 1994), or are imminent, but others remain far more optimistic as to the ability of modern societies to utilise and develop technology in such a way as to push back 'natural' limits to social advancement (Lomborg 2001; England 2000; Beckerman 1995 and 1994). The invocation of natural capital strongly tends towards 'a moral rejection of the view that humans can overcome nature's limits with their ingenuity' (Akerman 2003: 438). This overarching perspective on modern society is the key contextual factor in the debate under examination here – those invoking the importance of natural capital are often intensely critical of the impact of modern societies and development upon the environment and, indeed, upon aspects of human welfare intrinsically linked to the natural environment and natural processes. It is argued in this chapter that ecotourism, as advocated in the case studies, has a strong emphasis on development through the non-use of natural capital, and that its advocacy shares the 'moral rejection . . .' of development referred to above.

An example of this outlook is a forthright paper by Gowdy, who invokes natural capital in arguing for strong sustainability (Gowdy 1994). This author goes as far as to argue that 'de-development', rather than development, is necessary for *social* development, such are the limits to human advancement (ibid.). In justifying this view, he cites 'co-evolution' between the human race and ecosystems – both have evolved in a relationship to one another, a relationship that had in the past enabled important natural processes to coexist and co-evolve in a relatively harmonious way. This relationship, he argues, was upset by the advent of agriculture thousands of years ago and, more recently, by the advent of industrial societies (ibid.). Remarkably, he argues for the *re-creation*, and preservation, of pre-agricultural environments, in order to redress the balance towards ecological processes and away from human determined processes (ibid.).

Specifically, Gowdy argues that the 'non-development' of natural capital can be justified in a modern context through its welfare benefits as a resource for leisure (ibid.). In Gowdy's view, leisure activities – ecotourism prominent among them – can provide possibilities to push forward an agenda that is, at a macro scale, in favour of 'de-development' and, at a micro scale, prepared to argue for development to be limited to that which can take place on the basis of the 'non-development' of natural capital.

Gowdy's unreserved and unqualified advocacy of strong sustainability is striking, and strikingly ecocentric. Yet the thinking behind ecotourism mirrors Gowdy's explicit assumption of the non-substitutability of human-created capital for natural capital.

What is meant by 'environmental fragility'?

Some environments may be regarded as being more 'fragile' than others. The conservation NGOs featured in the case studies, WWF and CI, see this fundamentally in terms of the need to avoid the disturbance of important areas of biological diversity (see Chapter 3). Biodiversity sustains itself within an ecosystem, and development may alter the ecosystem to the detriment of biodiversity. In this sense, natural environments can be regarded as intrinsically fragile in the face of development.

However, it is equally common for conceptualisations of fragility to make direct reference to *human* communities. To be precise, they often look at fragility in terms of the *relationship* between the natural environment and the community living within it.

For example, two authorities, Harrison and Price, writing on tourism in fragile environments, see them as including those that exhibit 'marked seasonality, which means that many human activities are limited to quite clearly defined parts of the year' (Harrison and Price 1996: 1). Such activities are listed as 'cultivating crops, collecting naturally growing foods, hunting, or fishing [which are] typically limited to relatively few months – or even weeks – of the year' (ibid.). These months or weeks may also be the seasons for tourism, hence the propensity for tourism to impinge upon the environment and the way the community utilises it.

Harrison also refers to 'fragile lands', which he argues is similar to the conception of fragile environments utilised at the UN Conference on Environment and Development in 1992, and to that of another author, Deneven, writing about environmental fragility in Latin America (Harrison and Price 1996: 3–4). Deneven emphasises that such fragile lands should be managed according to traditional land use systems (cited in Harrison and Price 1996: 4). In essence, Deneven argues that these fragile environments impose specific limits to development on their respective communities. For its advocates, ecotourism has the advantage that it can sustain something approaching these traditional land use systems, yet also, potentially, deliver limited economic development based upon this.

Moreover, Deneven argues that 'social fragility, in terms of organisation, markets, prices, incomes, social relationships and politics [. . .] can be more critical than environmental fragility' (cited in Harrison and Price 1996: 4). Development is problematic not simply in relation to the environment, but also in relation to the existing balance *between people and the environment within a particular area.* On similar lines, Cater refers to the potential for even the most sensitive of tourism developments to bring problems to '*delicately balanced* physical and cultural environments' (Cater 1992: 19; my italics). This is a crucial point in the environmental critique of tourism development – environmental preservation is justified not just for its own sake, but with regard to the existing relationship between people and nature within defined local areas. Hence, the environmental critique of tourism is at one and the same time a critique of its cultural effects, with culture constituted as this relationship.

This approach is not specific to the advocacy of ecotourism. It is influential generally in the advocacy of sustainable development. For example, the UN's *Agenda 21* documentation, arising from their 1992 Conference on Environment and Development, asserts that:

> [indigenous people] have developed over many generations a holistic traditional scientific knowledge of their lands, natural resources and environment. [. . .] In view of the interrelationship between the natural environment and its sustainable development and the cultural, social and physical well-being of indigenous people, national and international efforts to implement environmentally sound and sustainable development should recognise, accommodate, promote and strengthen the role of indigenous people and their communities.
>
> (UN 1993)

The quotation makes explicit that it is the *relationship* between local (in this case indigenous) communities and the natural environment on which they rely that is central to sustainable development, and it is this relationship that is deemed fragile in the face of potentially transformative development.

The discussion of fragile environments, then, *extends* the notion of fragility from specific natural environments to the local communities that inhabit them. In this vein, the relationship between fragile environments and fragile communities has been described in the following way:

> Just as traditional uses of soils, waters, plants and animals – often developed over centuries (or longer) of experimentation to minimise change in communities' biophysical life-support systems – may be rapidly degraded by external influences, the communities' societal structures are equally susceptible to change by external human forces, whose magnitude and potential impacts are not always predictable.
>
> (Harrison and Price 1996: 5)

As a description of the way rural communities relate to their natural environment, the above quotation is uncontentious. However, as a justification for the maintenance of that relationship, it is open to question.

The case studies, environmental fragility and natural capital

A different approach to summarising the data from the case studies is taken in this chapter. The data is developed in a logical, thematic order, rather than case study by case study.

An emphasis on the non-consumption of natural capital

There is a clear emphasis on the non-consumption of natural capital in the advocacy of ecotourism. This is central to the general rationale for ecotourism ICDPs as advocated by a range of authors (Fennell 2003; McClaren 1998; Boo 1990; Ziffer 1989). This rationale is that ecotourism can constitute sustainable development in the rural developing world on the basis that it can bring a symbiotic, or mutually reinforcing, relationship between conservation and development, two concepts normally regarded as contradictory.

The IYE is an important point of reference here, as it involved a broad range of conservation and development oriented NGOs, and produced the influential *Quebec Declaration* and *The Final Report* (UNEP/WTO 2002a and 2002b) based on this. These documents make direct reference to the 'symbiosis' argument. For example, the *Final Report* argues that ecotourism should be about how communities 'both conserve and derive benefits from natural and cultural resources' (UNEP/WTO 2002a: 82). More specifically, the IYE presents living off the non-consumption of natural capital in a positive light, even suggesting that a whole continent could prioritise this in its development outlook:

> Conservation of natural resources can become *mainstream to socio-economic development in Africa*. National parks and reserves in Africa should be considered as a basis for regional development, involving communities living within and adjacent to them. Given their strong international recognition, parks and reserves can be turned in to sort of brands, providing advantages in tourism marketing and promotion.
>
> (UNEP/WTO 2002a: 12; my italics)

Here, the document argues that Africa – the poorest of the continents – can major in conserving natural capital, rather than through the transformation of nature in pursuit of development.

An imperative to preserve fragile ecology, or to sustain an economic asset?

The conservation NGOs have environmental conservation at the centre of their agenda, and hence, in the first instance, are concerned with the conservation of natural capital for environmental ends. WWF rationalises this through the concept of 'ecoregions', or 'ecologically fragile regions', which refer to specific environments that it deems particularly valuable, and particularly fragile in the face of development (WWF-International undated d). These regions are deemed to comprise ecosystems containing important biodiversity. There are 238 such regions, almost exclusively in the developing world (WWF-International undated e). Biodiversity tends to be richer in these, in large measure by virtue of the lack of development itself.

WWF refers to the need to 'reconcile human development needs with those of biodiversity conservation within large-scale areas' with the aim of 'ecoregion conservation' (WWF-International 2001c: 3). These large-scale areas are characterised by relatively undisturbed natural environments, inhabited by small-scale, rural, economically poor, agrarian, human communities. WWF seeks to influence the way 'natural resources and the environment are used and changed by people' in these regions (ibid.), and its interest in ecotourism stems from this. Further, '[w]here tourism is a major activity in an ecoregion, it is important that the conservation vision and strategy for that ecoregion takes account of the threats and opportunities posed by tourism' (ibid.).

WWF notes the coincidence of biodiversity with natural attractions suitable for tourism. It refers to the fact that tourism developments often occur in 'environmentally fragile areas that are biologically significant and rich in wildlife' (WWF-International 2001a: 1). It also refers directly to 'fragile regions' in its literature on tourism. In its *Tourism Background Paper* it argues that:

> [g]iven the ecologically fragile regions, such as those that include coastal areas and coral reefs, are often attractive as tourist destinations, inappropriate or unplanned development can be disastrous in terms of habitat degradation and biodiversity loss, and can result in the misuse of natural resources such as freshwater, forests and coral reefs.
>
> (WWF-International 2001c: 1)

In contrast to this, ecotourism is deemed to be not just a more benign form of development, but specifically a more benign form of tourism, too (ibid.).

Similar themes are echoed in the other conservation organisation among the case studies. CI utilises a concept very similar to WWF's 'ecoregions' – that of 'biodiversity hotspots' (CI undated i). Its interest in tourism and development arises principally from its core aim of preservation with regard to these biodiversity hotspots (CI/Christ *et al.* 2003). Biodiversity hotspots are those parts of the world that contain the richest biological diversity (CI undated i). The majority of these hotspots are in the less developed or middle-income countries (see list at CI undated g). It is here that ecotourism ICDPs are a tool in CI's armoury for achieving conservation and development.

SNV, as a development NGO, does not have a developed conception of environmental fragility as being rooted in environmental imperatives in the way that the conservation organisations do. However, it tends to view the fragility of the environment as mediated through the relationship of the community to that environment. This mirrors the point made by a number of authors (e.g. Harrison and Price 1996; Cater 1992), alluded to in the earlier discussion of environmental fragility, that it is the relationship between human activities and the environment that is key, rather than the environment in and of itself.

This linking of environmental fragility to the relationship between people and nature is illustrated in SNV's discussion of carrying capacity in its literature. SNV invokes the notion of a carrying capacity, 'refer[ring] to the possibility of the area to support tourism development' (SNV/Caalders and Cottrell 2001: 31). However, how it is conceptualised and even calculated inevitably involves judgements about the relationship between the community and the natural environment. SNV has devised the following checklist in order to help 'estimate how many tourists could visit the area without causing negative impacts on the local culture and environment' (SNV/de Jong 1999: 30):

- the size of the area where the tourism product is organised;
- the degree to which local people have been exposed to the outside world and to tourism;
- the facilities for tourists in the area;
- the number of local people who can provide services (especially tours) to tourists;
- the ecological vulnerability of the area.

Here, ecological fragility is involved, but alongside factors relating to the prior level of development and culture, or, in other words, the relationship between the community and its environment. It accepts that, while carrying capacity should not be exceeded, 'tourism development will cause certain changes' and it is therefore 'important to determine with the community the accepted amount of change' (SNV/Caalders and Cottrell 2001: 31).

That is not to say that ecological imperatives are not also prominent for SNV – a distinctly ecological carrying capacity is also recognised in its literature. For example, in Africa, too many tourists are mooted as a problem in relation to 'the fragility of the attractive ecosystems, such as the Okavango, the Makgadikgadi pans and the major rivers and adjoining forests' (SNV/ Rozenmeijer 2001: 13), as it is held that, '[t]hese ecosystems cannot cope with large tourist numbers' (ibid.). And although tourism can be of benefit, it can also be a problem when it threatens the 'disturbance of ecosystems' (SNV/Caalders and Cottrell 2001: 11).

Elsewhere, SNV states that '[t]he natural resources in and around the area should be protected (in order to preserve the natural attractions), and the tourism activities may not have a negative effect on the environment' (SNV/ de Jong 1999: 28). Here, that development is to be limited to that which local environmental conservation will permit, is explicit, and justified with reference to the environment's role as an income earner for the community as a natural attraction through its non-consumption.

One example of where carrying capacity has influenced SNV's operations is with regard to ecotourism in Botswana. SNV is involved in a number of community-based ecotourism projects here (SNV/Rozenmeijer 2001). In these projects there is a conscious 'low volume – high value' policy, partly to safeguard the 'exclusive "wilderness experience"' from mass tourism (ibid.: 13). Here, a strict carrying capacity is deemed to have a clear *commercial* argument in its favour, but also one coinciding with conservation of the natural environment. Further, in a section of a document addressing the marketing of ecotourism, SNV argues that, '[t]he tourism resource is the natural, cultural and socio-economic environment' (SNV/Caalders and Cottrell 2001: 38), and that these '[u]nique resources can be a national park, a specific animal (elephants or lions), indigenous culture, landmarks (Mount

Everest), unique buildings . . .' (ibid.). If the resource is the natural environment, its protection from too many tourists or too much development can be justified in terms of material benefits for the community rather than an environmental imperative per se. Yet at the same time these material benefits cannot extend beyond those available through the existing 'natural, cultural and socio-economic environment' (ibid.), which is, in this formulation, the 'tourist resource' (ibid.). Thus, the benefits may be regarded as limited in practice by an environmental imperative not dissimilar to that invoked by the conservation organisations.

Tourism Concern has made its reputation on the basis of their 'people first' stance – it has been explicitly critical of the 'nature first' priorities of some conservationists and ecotourism advocates. Indeed, Tourism Concern is clear on differentiating itself from conservation organisations (Barnett 2000). In an editorial in Tourism Concern's quarterly magazine *In Focus*, journalist and prominent Tourism Concern Council member Sue Wheat writes about the organisation frequently getting requests from the media to comment on tourism's effect on species and the environment. Her response is that Tourism Concern 'focuses more on the impact of tourism on people and their environment' (Wheat 1997: 3). Wheat cites the example of Burma to illustrate the different priorities between some conservation policies and Tourism Concern's outlook. She points out that 'when conservation of wildlife takes priority in order for tourism to be developed, the people who live there often suffer badly' (ibid.). She refers to the reports of murder and eviction in Burma in the process of creating wildlife reserves, reserves in which international wildlife agencies have been involved. Indeed, WWF has, in the past, been criticised for its alleged complicity in such activities in Namibia (Mowforth and Munt 1998: 176). The vital thing is, Wheat argues, to see 'the importance of local people and wildlife *co-existing*' (Wheat 1997: 3; my italics).

In another editorial, Barnett, points out that '[i]t has taken a lot of work over the years to get people to understand that "sustainable tourism" involves people as well as wildlife' (Barnett 1999). The emphasis on people is prominent in 'Community Tourism', pioneered by the organisation which, as the name suggests, is strongly community oriented and explicitly critical of any instance where conservation appears to be at the expense of local people. *The Community Tourism Guide* (Tourism Concern/Mann 2000) emphasises this conviction (e.g. p. 12 and pp. 26–7).

Hence, Tourism Concern's approach appears to be a much more anthropocentric one – its principal focus is the well-being of the community itself. This is an approach that cannot be seen as a product of a prior aim to conserve the environment. Tourism Concern has consistently, from its inception, stressed the centrality of community well-being to its campaigning work (Barnett 2000). However, it shares with the other NGOs featured in the case studies a strong emphasis on development through the non-consumption of natural capital in rural development (Tourism Concern/Mann 2000: 26–7 and elsewhere).

Ecotourism as the least worst option for the environment

The shared emphasis on the non-consumption of natural capital, be it through differing rationales, means that development becomes an issue of maximising well-being within very strict environmental limits. For example, WWF often presents ecotourism as the *least worst* form of development with regard to the usage of natural capital. It argues that 'tourism should be integrated into broader regional priorities' (WWF-International 2001c: 3), these priorities being biodiversity conservation. In pursuit of this core aim, ecotourism may be a more acceptable alternative form of development:

> In certain areas that are particularly ecologically fragile, any form of tourism development may be inappropriate. Tourism is more acceptable, however, where its potential negative impact is judged to be less than that which might result from alternative development strategies such as mining or logging, or where the development of part of an area for tourism allows the remainder to be conserved.
>
> (ibid.)

Further: 'It should be planned, managed and undertaken in a way that avoids damage to biodiversity, and that is environmentally sustainable, economically viable and socially equitable' (WWF-International 2001a: 2). The document goes on to argue that '[t]ourism [. . .] should be undertaken [. . .] in preference to other potentially more damaging forms of development' (ibid.).

The potential economic benefits of ecotourism are obviously important for local communities. However, they are rationalised here on the basis that they may offset demands to utilise the environment in other ways deemed less sustainable.

It is accepted, however, that these damaging alternatives may be attractive to poor, rural communities. In a document on tourism's role in the conservation of large carnivores, WWF argues that obstacles to community involvement in projects might include '[p]ressure for more rapid economic growth' and also '[c]onflicting aspirations of local farmers and hunters with the emerging tourism industry' (WWF-UK 2000: 11). Here, there is a recognition that communities, or sections of communities, may favour less 'sustainable' options, and here tourism is vital in its ability to give wildlife and the environment an economic value through its conservation.

WWF tourism expert Justin Woolford backs up this view of tourism's role:

> In many of our field projects that we support, we are constantly looking for alternative sustainable livelihoods, and tourism is always one that comes up as something that people or communities can engage in that isn't going to be as environmentally damaging as potentially some other activities, like logging for example.
>
> (Woolford 2002)

This view is mirrored in one document as follows:

> Tourism is an important sustainable livelihood option for local communities dependent upon natural resources in many areas in which WWF supports projects. It can bring money and employment to areas previously engaged in unsustainable activities such as logging.
>
> (WWF-International 2001a: 2)

Here, it is argued that tourism may be the best, or least worst, option for conservation, given that the community requires a livelihood, and that on these grounds it is 'sustainable'.

CI also mirrors the view that ecotourism is justifiable on the basis that it is the least worst development option in terms of the effect on biodiversity, arguing that '[w]orking with communities to develop products and to open markets creates an economic incentive for them to conserve their natural resources rather than destroy habitats for farming, cattle ranching or timber extraction' (CI undated c).

To this end, CI is involved in major projects to incentivise developing world states, such as Gabon (in this instance working with WWF), to redirect economic activity away from activities consumptive of natural capital to non-consumptive ecotourism (CI/Christ 2004).

The argument that ecotourism can be the best, or least environmentally impacting, form of development also finds favour at SNV: 'Tourism may be less damaging to nature compared to alternative economic sectors such as agriculture and forestry' (SNV/Caalders and Cottrell 2001: 11). As such, encouraging ecotourism may provide an incentive for communities to engage in activities deemed more sustainable, based more closely on conservation of natural capital. SNV views positive impacts from its projects as including 're-valuation of ecological values by the local population and authorities as a result of tourism interest, as well as economic justification and means for protection of nature' (ibid.).

This can be true beyond the community, at government level too. For example, SNV tourism officer Marcel Leijzer makes the point that 'the tourism sector puts pressure on governments to combat cutting trees and become more serious about nature conservation' (cited in SNV/de Jong 1999: 27). He adds that '[o]ur guides take more notice of such things as well, after all, it is in their interests' (ibid.). He cites the following experience of being approached by a Finnish development organisation: 'They wanted a certain forest to attain international status and asked us to bring tourists to the area. When a forest has become a tourist attraction, it becomes easier to garner support for nature conservation' (ibid.).

As Tourism Concern is not directly involved in operationalising ecotourism, seeing its role more in terms of campaigning for social justice in relation to all sorts of tourism developments, it is not surprising that its literature is less specific on this point. However, it also strongly hints at the view that ecotourism may be positive in that it can be the *least worst* development option for the environment, as opposed to the option that yields the *greatest* level of development (notwithstanding different ways that development may be conceived). For example, it is made clear that community tourism's role is to provide an *alternative* development option in the face of pressures from logging, mining and other extractive industries (Tourism Concern/Mann 2000: 27). Further, '[i]f conservationists want [local communities] to say no to harmful development, they must offer them an alternative means of feeding their families. Tourism may be that alternative' (ibid.). There is no suggestion here that community tourism yields *optimal* development, just that it can yield *some* development premised upon leaving natural capital in a near pristine state.

Elsewhere, wider infrastructural development is condemned on the grounds that 'the best defence of many "unspoilt" wilderness regions has been their inaccessibility', and that '[n]ew infrastructure such as roads or airstrips opens up regions for incoming colonists and other destructive activities, such as logging and farming' (Tourism Concern/Mann 2000: 14). Community tourism, then, can be a bastion against 'logging and farming', and other activities potentially more lucrative than small-scale ecotourism, and is extolled as 'sustainable' for this reason.

Summary

In summary, there is a clear emphasis on the non-consumption of natural capital in the advocacy of ecotourism as sustainable development, and this is emphasised in all the case studies.

However, the conservation of natural capital can be justified in two distinctive ways. First, as an environmental imperative – certain ecosystems containing important biodiversity may be deemed important in and of themselves, or with regard to global environmental concerns such as the role of forests as 'carbon sinks'. Second, natural capital can be seen as worthy of conservation on the basis that it is an economic resource for the local community through its non-consumption. The conservation NGOs have a developed view of the first justification, while SNV, and especially Tourism Concern, emphasise the second. However, both views are common across the case studies, and are often expressed in terms of ecotourism being the least worst option for the environment that can deliver limited economic benefits.

Yet, to sustain the limited economic benefits of ecotourism, conservation of the environment has to be a priority, and hence the two strands of thinking are not as different as they might at first appear – ecocentric and anthropocentric organisations, and lines of thinking, are making essentially the same case from different starting points. Although one view could be characterised as 'environment first' and the other as 'people first' – two views ostensibly at odds with each other – this apparent tension masks a substantially common approach. Be it in its role as a local economic resource, or as a global environmental imperative, natural capital is to be preserved. Conservation and development priorities have been drawn together conceptually around *strong* sustainability, a *particular* view of sustainable development, considered below.

Indeed, the convergence here is a microcosm of the more general convergence between conservation and development in a significant strand of thinking on each, a convergence characterised as sustainable development, and discussed in Chapter 2. There it was established that conservation organisations have taken on board the need to combine conservation with development, and development thinking has taken on board the increased emphasis on conservation. Ecotourism's ability to integrate conservation and development was established as one expression of this trend.

What type of sustainability?

There are very different views on what constitutes a sustainable relationship between natural and 'human created' capital. What originally Beckerman (1994) termed 'weak sustainability' involves a recognition that natural capital values can be run down if human created capital is adequate to compensate for this (Adams 2001: 117–121; Beckerman 1995 and 1994). It allows for a dynamic relationship between human development and natural resources, and for the notion that resources can progressively be uncovered and better utilised precisely through development premised on using up natural capital (ibid.). 'Strong' sustainability, on the other hand, sees a pressing need to maintain stocks of natural capital, taking the relationship between development and the environment as being much more antagonistic (ibid.). Ecotourism errs towards a strong version of sustainability thus defined.

All of the case studies clearly take a strong sustainability stance in the rural areas in which ecotourism ICDPs are applied. They locate an economic value as rooted in the natural resource itself, and hence a value that can only be realised by leaving it as it is. If the resource is transformed, or destroyed, any aesthetic, scientific, spiritual and ecological value (all of which have been cited by ecotourism's many advocates – see for example Fennell [2003] or McClaren [1998]) of the resource is lost. So, too, will be the prospect of developing ecotourism commercially, and economic benefits arising from this. So ecotourism appears to offer us the best of both worlds – a strong sustainability promoting conservation, and some economic development based on this.

However, although the case studies suggest that they can combine a strong orientation towards environmental conservation with economic development, 'strong sustainability' has been criticised on the basis that it is implicitly *anti-development* in general (Beckerman 1995 and 1994) and also specifically with regard to tourism (Butcher 2006b). Beckerman's critique of strong sustainability is that basing development on natural capital in this way dictates how far and in what direction a community can progress. Any development that is transformative of the relationship between the community and their natural environment is ruled out of order, or 'unsustainable'. Beckerman describes this limitation as 'morally repugnant', as it 'impl[ies] that all other components of welfare are to be sacrificed in the interests of preserving the environment in exactly the form it happens to be in today' (Beckerman 1994: 192). What Beckerman is criticising is essentially the philosophy underlying ecotourism.

Critical natural capital and development

While an emphasis on development through the non-use of natural capital can be criticised for limiting economic development in poor, rural communities (Beckerman 1995 and 1994), many environmentalists would argue that ultimately the preservation of important ecosystems is so important that it should take precedence over development. Indeed, this is a point of view articulated by some conservationists, some of whom oppose ICDPs on the grounds that they compromise conservation (Oates 1999; Barrett and Arcese 1995).

It has been argued that a key issue is *critical* natural capital, referring to certain natural resources that cannot be replaced if lost, have no substitutes, and cannot be created or compensated for elsewhere (Buckley 1995). For the conservation oriented advocates of ecotourism, WWF's 'ecoregions' or CI's 'biodiversity hotspots' could be conceived in this way – areas of the globe with a high concentration of biodiversity that may be quite unique. A logical argument would be that the biodiversity contains important potential for the scientific understanding of nature, or perhaps contributes to the absorption of carbon emissions, to the extent that it is simply irreplaceable.[2]

But in such cases why should development for *people* be tied to these areas of important biodiversity? Why not focus on providing better prospects for communities away from such areas, where they can enjoy some of the advantages of modern development? The advocacy of ecotourism as sustainable development draws together conservation and development not just in theory, but also *spatially* – they must take place *in the same place at the same time*. This is essentially the 'symbiosis argument' (Goodwin 2000, Budowski 1976), the central argument behind the claims that ecotourism can constitute exemplary sustainable development. Given that ecotourism ICDPs adopt a strong sustainability approach, this argument would seem to tie development prospects for communities to severe, localised, natural limits.

Such a vision of development is very different from the experiences of every developed country in the world, where development has been premised upon urbanisation and a separation between people and the land through economic growth and the establishment of a division of labour. If for no other reason than this, the vision should be questioned. In fact, although it cannot be developed substantially here, this vision of development resonates with the colonial view of Africa – simply, that it was a continent incapable of development as experienced in the West (Leech 2002; Reader 1998). For example, the question of how to encourage ways of living that would enable Africans to live without transforming their natural environment was posed by scientist Barton Worthington, writing in 1938 on the subject of 'Science in Africa':

> [A] key problem was how Homo Sapiens could himself benefit from this vast ecological complex which was Africa, how he could live and multiply on the income of the natural resources without destroying their capital . . . and how he could conserve the values of Africa for future generations, not only the economic values but also the scientific and ethical values.
>
> (cited in Adams 2001: 37)

At this time, such a view was certainly part of the colonial outlook, albeit an insecure one, in Britain. This outlook viewed Africa as a place characterised by wilderness and less civilised races. Yet re-read the quotation – it would not be out of place in the *Quebec Declaration* (UNEP/WTO 2002b), especially given the reference to 'ethical values' and 'future generations', along with the cultural relativism implicit in the reference to 'the values of Africa'.

In the post-colonial period, too, one writer describes Western environmentalists as trying to persuade people in the developing world of 'the virtue of living off the income of their natural resources, not the capital' around the time of the 1972 UN Conference on the Human Environment in Stockholm (McCormick 1995: 49). Yet here, aspirant developing world nations, buoyed by their newly gained independence, sought to shed developing world status through ambitious development plans, plans that some environmentalists considered dangerous from the perspective of environmental conservation. The contemporary view emanating from the IYE, that '[c]onservation of natural resources can become mainstream to socio-economic development' (UNEP/WTO 2002a: 12) was often seen as restricting much needed economic development at this time (Adams 2001; McCormick 1995).

Of course, the discussion of natural capital and development today is framed very differently from the beginning of the twentieth century, the 1950s or even the 1970s. Today, post-colonial ambitions for development have been numbed by failed development projects, debt and dependency. Environmental conservation has emerged as a central concern in Western societies, and has a significant effect on development policies through the greening of aid (Adams 2001). The neopopulist critique of modern development, influential in rural development, has come to frown on thoroughgoing development in favour of small-scale, community-based initiatives that draw on local resources in a 'sustainable' fashion (e.g. Chambers 1997, 1988 and 1983).

However, there is an important continuity between the colonial view of inferior subjects, and the upbeat pronouncements of sustainable development through ecotourism – neither holds out much prospect of the poor nations becoming rich.

Conclusion

Ecotourism brings conservation and development together in *theory* and *spatially* around a strong sustainability stance. It allows little scope for the substitution of human created capital for natural capital in the rural communities concerned. Yet, as is evident throughout the study, this stance is readily associated with sustainable development. In this respect there must be a strong case for making explicit that ecotourism conforms to a *particular* and *contested* version of sustainable development, that of strong sustainability. This would potentially open up the discussion of rural development associated with ecotourism by compelling its advocates to address the development implications that are associated with strong sustainability more directly. The rubric sustainable development may in this respect mask a clear emphasis on conservation over development running through both the ecocentric and formally anthropocentric arguments for ecotourism from a diverse variety of NGOs.

Concerns over critical natural capital lead the conservation NGOs to the conclusion that communities' development should be based around its non-consumption. This, at a stroke, rules out transformative development. A more creative way of restating the problem may be to give ground to a conception of development as involving a separation between people and environment, an approach that runs counter to environmentalism (Pepper 1996). Yet the more systematic development that this would entail is eschewed in all the case studies featured. In the context of meagre aid budgets, and the relative dislocation of parts of the developing world from the world economy (principally in Africa), this emphasis on natural capital may be posed as pragmatic, or as a stepping stone to greater development. In general, it is presented as sustainable development, as a normative option, favoured above more thoroughgoing development options.

Thus, the strong sustainability emphasis on the non-substitutability of natural capital enables ecotourism to be presented as an innovative development option. Yet the strong sustainability assumption *itself* sets prior, severe limits on the prospect for development.

7 Symbiosis revisited

Introduction

Throughout the study it has been argued that a symbiosis between development and conservation is the central rationale for ecotourism as sustainable development in the rural developing world. Individual chapters have attempted to deconstruct this 'symbiosis' argument by looking at a number of sub-themes that are integral to this central claim, drawing evidence from the advocacy of ecotourism in the case studies and in the wider literature. These chapters looked at community participation (Chapter 4), the emphasis on traditional culture (Chapter 5), and finally the promotion of development on the basis of the non-consumption of natural capital (Chapter 6).

The main purpose of this chapter is to reformulate 'symbiosis' in the light of the criticisms developed in Chapters 4–6. In order to do this, the chapter looks at accounts of specific projects run by WWF, CI and SNV to illustrate the way they conceptualise symbiosis. In the case of Tourism Concern, and with regard to the IYE, the section draws upon exemplary references in their respective literatures, to the same end (neither of these two case studies is directly involved with putting ecotourism into practice).

Finally, the chapter develops a number of points arising from the summaries, and from the general emphasis on a symbiosis between conservation and development in the advocacy of ecotourism as sustainable development in the rural developing world.

'Symbiosis' in the case studies

WWF and the symbiosis between conservation and development: the example of Namibia

The background to the development of WWF's ecotourism ICDPs in Namibia is drought and poverty – a major drought in the early 1980s contributed to widespread poaching which threatened animal populations, most notably the rare black rhino (WWF-UK 1999: 1). Although it is accepted that, for the semi-nomadic inhabitants of the Kunene region in north-west Namibia, '[s]urvival is precarious' (ibid.), it was the survival of the rhino that prompted WWF's intervention.

After 'extensive lobbying' of the post-independence Namibian government by NGOs (WWF-UK 1999: 2), 'Communal Area Conservancies' were formed, within which communities are able to benefit financially from managing the stock of wildlife (ibid.). The conservancies have to meet 'stringent criteria' relating to conservation (WWF-International/Denman 2001: 7). They are very significant in rural Namibia. One source states that at the start of 2001, 7.5 million hectares of Namibia's communal area was under, or developing as, conservancies (Flintan 2001: 1).

WWF has, since 1992, led a consortium of national and international organisations in the implementation of a project based on the conservancies termed 'Living in a Finite Environment' (LIFE) (WWF-International/Denman 2001: 7). LIFE supports ecotourism ICDPs, seeing them as part of 'community based natural resources management' (WWF-International/Denman 2001: 7). LIFE was initially funded by USAID through WWF-US, although key partners have subsequently included WWF-UK and WWF-International (Flintan 2001: 1). The local NGO partner in Namibia is called Integrated Rural Development for Nature Conservation (IRDNC). WWF instigated IRDNC, and has been a key funder throughout its existence (Flintan 2001: 1).

IRDNC's goal is 'to link conservation and sustainable use of wildlife and other natural resources to the social and economic development of rural communities in Namibia' (cited in Flintan 2001: 2). Its objectives are as follows:

> To contribute to building up the natural resource base, *as the foundation of all development*, in communal areas; to develop the capacity of local communities to jointly manage with government the wildlife and other natural resources in communal areas; to facilitate the return of social and economic benefits from wildlife and other natural resources to the residents of communal areas; to promote community based natural resource management both nationally and internationally.
>
> (IRDNC cited in Flintan 2001: 2; my italics).

The conservancies can claim success in meeting these objectives. WWF points to 'encouraging signs that this integrated management of tourism and conservation is benefiting biodiversity', reporting that '[w]ildlife numbers, including black rhino and elephant, have increased significantly since the community approach has been adopted' (WWF-International/Denman 2001: 7). Also, ecotourism ICDPs in Namibia are argued to have helped to 'creat[e] alternative livelihoods for a rural, highly marginalised community' (WWF-UK 1999: 1), thus offsetting the need for less environmentally benign livelihoods (ibid.).

Overall, WWF argues that the conservancies are 'an exciting development – empowering poor, disenfranchised rural people, providing alternative livelihoods to their subsistence farming and conserving wildlife into the bargain' (WWF-UK 1999: 2).

The example of ecotourism in the nature conservancies of Namibia draws together the themes considered in this study. First, natural capital – previously important for communities as bushmeat in times of famine, as a resource for household construction and tools, and as a cash earner through trade (Flintan 2001: 1) – has now gained a value through *non-use*. This value may offset alternatives deemed unsustainable by WWF (WWF-International/Denman 2001: 7).

Also, the communities participate in the process, based on 'WWF's belief that giving people control over their own natural resources is the best way to ensure a thriving environment' (WWF-UK 1999: 2). Yet the process itself is subject to the financial authority of the NGOs and their funders, who hold the key to important, scarce rural aid funding. The local partner, IRDNC, would not exist were it not for the foreign NGOs.

Through the rationale outlined here, the development outlook for the communities is aligned to the project of conserving the local, natural environment – a symbiosis between conservation and development possibilities at the local level is achieved.

Finally, the project is argued to be exemplary sustainable development (WWF-International/Denman 2001), a theme running through all of WWF's advocacy of ecotourism.

CI and the symbiosis between conservation and development: the example of Botswana

One of CI's keynote projects relates to the San bushmen of the Okavango Delta in Botswana, a people who, according to CI, 'have lived in harmony with their natural environment' for '[t]housands of years' (CI undated, h: 1). There is a striking emphasis on the notion of 'harmony' between people and nature thus described. For example, the tours offer 'the rare opportunity to experience the traditional activities of the Bukakwe, and explore their intimate connection to the ecosystems of the Okavango' (ibid.). Further, the project has created jobs that 'allow the Bukakwe to keep their ancient heritage alive while protecting the environment' (ibid.). These jobs include craft making, traditional dance, storytelling, spear throwing, traditional food tasting and other activities that attract the tourist on the basis of tradition (ibid.: 1–2).

Economic benefits are directly linked to the supporting of tradition. For example, one project, the Gudigwa camp, offers the community 'economic opportunities' to 'revive their traditional way of life' (ibid: 1). This is true for conservation, too. CI and its partners 'hope to create economic opportunities and incentives for the community to adopt land use practices that will protect the health of the local ecosystem' (ibid.), as well as to realign land use to enable a wildlife corridor to be established.

The project is presented as a product of extensive dialogue between CI and the community, and is 'fully owned' by the community trust established through this dialogue (ibid.). However, the extent of 'ownership' is not explored.

The San communities of western Botswana are among the poorest on the planet. They remain marginalised within Botswana, and isolated economically. The traditions of the San revolve around what could be described as an intimate and fragile relationship with nature (in the sense discussed in Chapter 6), and it is certainly feasible that tourism can yield limited welfare benefits. However, quite explicitly in this example, it seeks to do this on the basis of *reinforcing* rather than challenging this relationship with the natural world. The extent to which this could set in motion wider development is doubtful, and indeed such development is not an aim of the project – clearly, wider economic development would be problematic from the perspective of conservation.

The project reflects the themes of this study. CI argues that it involves community participation, yet the terms of the project itself were settled prior to this process. These terms are a strong orientation towards traditional aspects of the society, and a non-negotiable stewardship role for the local community with regard to its natural surroundings.

SNV and the symbiosis between conservation and development: the example of Botswana

SNV is also heavily involved in Botswana. Its ecotourism ICDPs are concentrated in the west of the country. This area has experienced relatively little economic development, even though the country has performed relatively well in the region in economic terms. In the Kalahari region SNV argues that there are relatively few natural resources other than the wildlife and veld products (these are fruit, berries, tubers and leaves), and these resources lie mainly in protected areas and hunting zones (SNV/Rozenmeijer 2001: 7). The former are controlled mainly by the state, the latter communally, but in practice, with little management (ibid.). In the 1990s community-based natural resource management was developed by SNV. This tries to encourage specific local communities to manage their natural resources in a manner that benefits them economically and that conserves the wildlife. One document introduces the case for ecotourism ICDPs clearly:

> In Botswana the focus is on CBNRM. The idea behind this approach is that when communities realise the economic value of their surrounding natural resources, they are inclined to manage them in a more sustainable way. The aim of this approach is twofold: to create rural economic development and to conserve natural resources.
>
> (SNV/Caalders and Cottrell 2001: 22–3)

Here, development is proposed on the basis of the community managing its natural resources through CBNRM, in a way that ensures the conservation of these resources.

As at 2001, about fifty community organisations were involved in SNV's CBNRM projects all over Botswana, based around trophy hunting, photography and nature-based safaris, overnight accommodation and self-drive, culture and handicrafts (SNV/Rozenmeijer 2001: 10–11). The aim here is to promote the economies of these areas around non-consumptive, traditional industries, and this is described as 'sustainable tourism development' (SNV/Rozenmeijer 2001: 5; also see SNV/Caalders and Cottrell 2001: 4–5).

It is worth noting that 22 per cent of Botswana's land mass is designated as wildlife management areas (SNV/Rozenmeijer 2001: 8) – it is in these areas that development is promoted on the basis of the non-consumption, or non-use, of natural resources. SNV's approach, thus, applies potentially to large swathes of the country and very many rural inhabitants.

The benefits of the approach adopted by SNV are listed as: income and employment; the adding of value to the national tourism product; and most notably that 'the benefits derived from the use of natural resources for tourism will prompt the community to use these valuable resources in a sustainable way' (SNV/Rozenmeijer 2001: 13).

A further list of more specific advantages is given: aid for CBNRM incentivises conservation for governments; does not require heavy investment (being based on natural resources); justifies the allocation of natural resources from government to community; 'enhances the value of culture'; 'enhances the value of and pride in the natural environment'; 'encourages a sustainable management of the environment'; and 'the "sustainable use of the environment" dimension of CBT helps sell the idea of NGO assistance to financers' (SNV/Rozenmeijer 2001: 14).

The project is exemplary of all the themes developed thus far in the study. It offers development based on the preservation of natural capital. It involves the community directly in the management of its natural resources to this end, and incentivises this. Ecotourism, it is held, can enhance the value of traditional culture, thus helping to conserve it. It does not require heavy investment, relying as it does on natural capital. However, the agenda summarised above – the basis of SNV's involvement in Botswana – is not subject to community participation. Finally, this type of development is referred to as 'sustainable development', and is in turn presented as a counter to forms of development implicitly regarded as unsustainable.

Tourism Concern and the symbiosis between conservation and development: an outlook central to 'community tourism'

Tourism Concern, as a campaigning organisation, does not operate ecotourism ICDPs. Yet its keynote publication on community-based ecotourism, *The Community Tourism Guide* (Tourism Concern/Mann 2000), includes a directory of holidays/projects featuring the projects mentioned above and many others that are funded by a range of conservation and development NGOs, including WWF, CI, SNV, the Audubon Society and Rainforest Concern.

Its 'Ten Principles for Community Tourism', elucidated in the *Guide*, are exemplary of its overall approach, and are listed in full below:

1. Community tourism should involve local people. That means they should participate in decision making and ownership, not just be paid a fee.
2. The local community should receive a fair share of profits from any tourism venture.
3. Tour operators should try to work with communities rather than individuals. Working with individuals can create divisions within a community. Where communities have representative organisations, these should be consulted and their decisions respected.
4. Tourism should be environmentally sustainable. Local people must benefit and be consulted if conservation projects are to work. Tourism should not put extra pressure on scarce resources.
5. Tourism should support traditional cultures by showing respect for indigenous knowledge. Tourism can encourage people to value their own cultural heritage.
6. Operators should work with local people to minimise the harmful impacts of tourism.
7. Where appropriate, tour operators should keep groups small to minimise their cultural and environmental impact.
8. Tour operators or guides should brief tourists on what to expect and on appropriate behaviour before they arrive in a community. That should include how to dress, taking photos, respecting privacy.
9. Local people should be allowed to participate in tourism with dignity and self-respect. They should not be coerced into performing inappropriate ceremonies for tourists, etc.
10. People have the right to say no to tourism. Communities who reject tourism should be left alone.

(Tourism Concern/Mann 2000: 25)

On the face of it, the principles articulate no more than a broad aspiration for fairness, and for the community to be empowered. Yet for the NGO-funded projects advertised in the guide, NGOs are the drivers of development, and community participation is limited to their implementation, as argued in Chapter 4. Principle four suggests that these are, after all, conservation projects, although clearly there is the aspiration for the communities to derive the maximum possible benefit from conservation. Other principles, numbers 6 and 7, also suggest that transformative development is harmful. 'Traditional culture', 'indigenous knowledge' and 'cultural heritage' are also invoked to emphasise the importance of the community's way of life in shaping, and limiting, development. Indeed, principle seven indicates that any cultural impact should, 'where appropriate', be 'minimise[d]'.

Elsewhere, *The Community Tourism Guide* strongly promotes the argument for symbiosis between conservation and development. For example, the projects are lauded as a 'source of inspiration' to tourists on the basis that '[t]hey can show western visitors that a "sustainable lifestyle" and "living with nature" are practical realities, not just utopian concepts' (Tourism Concern/Mann 2000: 23). A harmony, or symbiosis, between the community and their environment is, then, a strong and recurring theme, mirroring the approach of the other case studies. In fact, the 'Ten Principles' closely reflect the general 'symbiosis' argument, and the assumptions underpinning it.

The UN IYE and the symbiosis between conservation and development: a view central to ecotourism's 'values and principles' (UNEP/WTO 2002a: 26)

The symbiosis between conservation and development is at the centre of the documentation from the UN IYE. The IYE is an important source in this respect – it reflects the views of organisations whose primary purpose is conservation, as well as those who are concerned in the first instance with well-being or development.

A number of excerpts from the IYE documentation illustrate this clearly. The section of *The World Ecotourism Summit Final Report,* reporting on its 'Asia-Pacific Forum', argues that '[m]ountainous areas often display a particular cultural richness, economic fragility, a decline in traditional populations and activities, and sensitive biodiversity. Mountain communities can use ecotourism to address these issues' (UNEP/WTO 2002a: 19). It is suggested here that ecotourism can tackle the conservation and well-being imperatives simultaneously. However, well-being is discussed in terms of 'traditional populations and activities', 'cultural richness' and 'economic fragility' (ibid.). Other types of economic development, based on the consumption of natural capital, may undermine 'traditional activities', and may upset 'sensitive biodiversity' (ibid.). The scope for economic development is constrained in this formulation by the importance assumed by conservation and tradition.

The link between economic development, in the form of tourism revenue, and conservation is a direct one – revenue is to be dedicated towards conservation thus: 'Financial and fiscal mechanisms should be implemented to ensure that a significant proportion of the income generated from ecotourism remains with the local community *and is reinvested for environmental and cultural conservation purposes*' (ibid.: 55; my italics). This seems to suggest that economic development is closely tied to a particular philosophy, one that sees natural capital and traditional culture as the principal sources of sustainable development in the rural developing world.

Elsewhere, a symbiosis between development and conservation is advocated in terms of a 'national vision of how ecotourism can serve biodiversity, as well as how biodiversity can serve ecotourism' (ibid.: 26). Here, the mutuality between the two – conservation and development – is explicit, as each is to benefit the other. Elsewhere, in similar vein, it is asserted that ecotourism should increase 'economic and social benefits for host communities, activity contributing to the conservation of natural resources and the cultural integrity of host communities . . .' (ibid.: 66), and this is equated with a 'sustainable' approach (UNEP/WTO 2002a). Again, it is clear that development gains should 'actively contribute' to conservation of natural capital. In the section on 'Guidelines to the Private Sector', the *Quebec Declaration* argues that operators need to 'conceive, develop and conduct their business minimising negative effects on, and positively contributing to, the conservation of sensitive ecosystems and the environment in general, and directly benefiting and including local and indigenous communities' (UNEP/WTO 2002b: 70). Here again, and throughout the document, conservation and development are referred to together, as symbiotic.

Although there is recognition that the development of ecotourism may compromise conservation, that *conservation* may place severe limits upon *development* is not considered anywhere within the lengthy documentation. The only possible exception to this is the acceptance that 'the assertion that developing ecotourism is a good method of solving the problem of poverty in developing countries should be expressed with caution' (WTO/UNEP 2002a: 53). However, this argument is put with reference to the potential for ecotourism to upset subsistence agriculture and traditional economic activities, rather than through a comparison with other, new development possibilities. It is an argument that compares funded ecotourism ICDPs to *what is*, rather than to *what could be*. It fails to conceptualise development outside of a pre-existing relationship between the community and its natural environment, and as such may constrain development thinking on regions that, as the document states, 'contain millions of people living in poverty' (ibid.: 43). For example, any consideration of the opportunity cost of ecotourism – other alternatives forgone through devoting scarce funding to ecotourism – is limited by the assumption that projects proposing thoroughgoing development do not constitute sustainable development.

The IYE documents refer to ecotourism's 'principles' and 'values' (e.g. UNEP/WTO 2002a: 26, and elsewhere). It is useful to see ecotourism in this way – this study has focused on establishing and critiquing the 'values' underlying the advocacy of ecotourism as sustainable development in the rural developing world. Central to these values, as presented in the documents, is a '*people centred* conservation approach' (ibid. 2002a: 43; my italics) in the world's biodiversity hotspots. Yet the well-being of people is closely integrated with conservation on a localised basis, as illustrated in the quotations above.

These values are also echoed in the view put forward in a section of the *Final Report* titled 'Working Group on Ecotourism Policy and Planning: the Sustainability Challenge', which states that '[h]umans should be recognised as being part of the ecosystem (as opposed to only using ecosystems)' (ibid.: 26). In one sense, it is a truism that humans are part of ecosystems – they exist in a relationship with the natural world in which changes to one part impact upon other parts in a systemic fashion (Hettne 1990: 183–6). However, in another sense, humans are distinctive within nature as having the capacity to harness and organise nature around distinctly human ends. Traditional definitions of development have been premised upon this supposition (Preston 1996: 118–19), and it also reflects the experience of economic development in the developed world (ibid.). The values that accompany the advocacy of ecotourism are distinctive, and seem to reject the modern legacy wholesale.

The IYE is the most important summary here, representing as it does a keynote UN conference involving the other NGOs featured alongside many others. It also reflects the principal themes examined in the study. Most notably, the symbiosis between conservation and development is a central theme running through all the documentation, underpinning the view that ecotourism plays a 'leadership role' in bringing about sustainable tourism development (UNEP/WTO 2002b: 67).

Symbiosis as the central theme

In addition to the above summaries, it is notable that all the case studies allude to the view that there is a symbiosis of sorts between the tourism industry *in general* and conservation. For example, WWF sees the tourism sector and itself as sharing a common goal, which is 'the long term preservation of the natural environment' (WWF-International 2001a: 1). CI shares this view, stating that '[p]erhaps more than any other sector, the tourism industry has a vested interest in protecting the natural and cultural resources of the areas upon which its business depends' (CI undated j).

The same point is made in *Beyond the Green Horizon: Principles of Sustainable Tourism* (Eber 1992), published jointly by Tourism Concern and WWF. SNV also alludes to this in its literature. For example, SNV's director Thea Fierens asserts that '[t]ourism projects often have close links to [. . .] natural resource management . . .' (SNV/Rozemeijer 2001: 5). Finally, the *Quebec Declaration* cites tourism's 'potential contribution to poverty alleviation and environmental protection in endangered ecosystems' as the reason for the industry's importance in achieving sustainable development (UNEP/WTO 2002b: 65). Having established this, the document then argues that ecotourism is exemplary in this respect (ibid.).

A symbiosis between tourism *in general* and the environment is also invoked in many discussions of the industry. For example, for Gunn (1987: 245), 'resource assets are so intimately intertwined with tourism that anything erosive to them is detrimental to tourism. Conversely, support of environmental causes, by and large, is support of tourism'. Writing from a conservation perspective, Croall makes a similar assertion in his apocalyptically titled *Preserve or Destroy? Tourism and the Environment* (1995), as does Neale in *The Green Travel Guide* (1998). In both publications, ecotourism is portrayed as a type of tourism that bases itself on this coincidence of interests in way other types of tourism, by implication less sustainable, cannot.

However, that tourism and conservation have a common goal, or that 'support of environmental causes', constitutes 'support of tourism' (Gunn 1987: 245), can only be true in the loosest sense. It only fully holds up with regard to small-scale ecotourism in small, rural communities – here there may be a coincidence between tourism and the prioritisation of conservation. The vast majority of tourism development impinges on the natural environment in a fashion that is clearly negative measured against WWF's aim to 'conserv[e] the world's biological diversity' (WWF-International 2001a: 1), or in, as the *Quebec Declaration* has it, the 'protection of endangered ecosystems' (UNEP/WTO 2002b: 65). WWF's 'vision for tourism' is that it 'should maintain or enhance biological and cultural diversity' (WWF-International 2001a: 1). The large-scale coastal developments so popular with tourists can hardly be said to do this (many of which are in regions that in the 1950s would have been essentially rural, relatively poor, and inhabited by small communities based around small-scale agriculture and fishing) and, indeed, WWF itself often cites such mainstream developments as negative with regards to biodiversity (notably, WWF-Mediterranean Programme 1999: 2).

The symbiosis between conservation and development, as formulated in all the case studies, is not characteristic of tourism in a general sense. Tourism is mainly from developed countries to other developed countries, and is frequently city based (WTO 2003). It relies on roads, airports, hotels, theatres, jet travel, hi-tech attractions and the internet. In the developed world it relies on, and feeds into, economic development. Yet in the rural developing world, lacking in modern infrastructure but rich in biodiversity, ecotourism projects feed into the maintenance of natural capital, eschewing modern development. Moreover, positioned under the rubric 'sustainable development', ecotourism is lauded as exemplary development. It is not, as has been evident throughout the study, seen as a first step towards greater developmental possibilities, or one of a number of options to be assessed on the basis of their capacity for improving the well-being of the community, but rather as a favoured sustainable solution for the rural societies concerned.

This suggests a different standard against which tourism development in the developing and developed worlds is judged – a symbiosis between development and the environment for those in the rural developing world, and a realisation of the economic benefits of transformative development in the developed world. In the latter case, it is less likely that conservation and development would be seen as symbiotic – rather they would be considered as important, but *competing*, priorities.

Institutional symbiosis

It has been shown that the rationales for ecotourism ICDPs from quite different NGOs coalesce around the symbiosis between development and conservation. However, an additional point worth noting is the way in which the symbiosis between conservation and development in *theory* is reflected in an *institutional* symbiosis. Organisations of both a development and well-being bent respectively, not only have a similar outlook in relation to ecotourism ICDPs, but they often work together in tandem to develop them.

There are many examples of this institutional symbiosis evident in the case studies featured. Most obviously, the UN IYE itself reflects institutional symbiosis, as it brought together both conservation and development organisations to share in the promotion of ecotourism as sustainable development. It is very much part of a post-Second World War trend in global conferences, organised through the UN, that progressively treat development and conservation together (Adams 2001: 54–79), bringing together individuals and organisations of both a conservation and development orientation respectively.

Also notable is the coming together of individual NGOs around projects and initiatives. For example, Tourism Concern and WWF collaborated on the publication of *Beyond the Green Horizon: Principles of Sustainable Tourism* (Eber 1992), a publication that set out the arguments for sustainable tourism. In part, its aim was to try to get tourism onto the agenda of the keynote UN Conference on Environment and Development held in 1992 (Barnett 2000). The two also collaborated in 1996 on another publication looking at tourism and sustainable development titled *Sustainable Tourism: Moving From Theory to Practice* (Forsythe 1996). Yet Tourism Concern has retained a critical stance on the record of conservation organisations in this field, and its director is adamant that it has never been, and should not be, referred to as a conservation organisation (Barnett 2000). Further, *The Community Tourism Guide* (Tourism Concern/Mann 2000) features large numbers of ecotourism ICDPs which are organised by conservation NGOs and premised upon their conservation aims.

SNV, while primarily a development organisation, works closely with conservation organisations on some of their projects too. For example, it has worked with WWF on projects financed by the Dutch government (SNV/Caalders and Cottrell 2001: 13). A further example of this is its work with the International Union for the Conservation of Nature (IUCN) and the IUCN-NL (the Dutch NGO members of the IUCN). The IUCN's focus is on the protection of nature and the sustainable use of natural resources, rather than development primarily (ibid.: 15). SNV works with IUCN on tourism projects in Vietnam and Botswana (ibid.). Indeed, SNV often refers to one of its goals as CBNRM when discussing tourism ICDPs (e.g. SNV/Rozenmeijer 2001: 7–11), which is also very much in keeping with the aims of the IUCN. CBNRM is presented as a means to development, but the potential for nature conservation to be a competing priority with development is unexamined (ibid.). SNV also contributes to IUCN projects whose explicit aim is nature conservation (SNV/Caalders and Cottrell 2001: 15).

Where the organisations are formally working together in this area we can, then, identify an institutional symbiosis. It suggests a merging together of organisations that were traditionally about development and conservation respectively, around the argument that the two aims can be symbiotic and therefore logically inseparable. Although not developed in this analysis, this is notable, and would seem to fit with the broader trajectory of aid to address 'green development' through the 'greening of development aid' (Adams 2001: 325–32).

Reconstructing symbiosis

The 'symbiosis' view can be reconstructed in the light of the criticism of its basic assumptions made in Chapters 4, 5 and 6. It was noted in Chapters 4 and 5 that all the case studies have a strong anthropocentric bent rhetorically. This is evident through their emphasis on participation and on traditional culture. In both cases, people, rather than nature, seem to be at the centre of the projects. People can not only benefit from development, but can also participate in the process (Chapter 4), and development can derive from *their* culture, reflecting the agency of the community itself (Chapter 5). This gives the advocacy of ecotourism a strong rhetorical anthropocentric flavour.

Yet culture is interpreted as *traditional* culture, in a fashion that elevates this above culture as embodying the desire and capacity to change (Chapter 5). Development, too, is based upon the non-consumption of natural capital, ruling out as unsustainable the *transformation* of nature for human ends (Chapter 6). The emphases on tradition and on natural capital are self-reinforcing – traditional human land use systems tend to have co-evolved in a relatively harmonious relationship with local biodiversity, unlike more modern agricultural practices or indeed modern industry (Gowdy 1994). Through this dual assumption – the importance of tradition and non-consumption of natural capital in development – the agency of the community is tied closely to a severe sense of natural limits to development. Hence, though the arguments for ecotourism prioritise culture, culture is closely tied to nature. The notion of 'biocultural diversity' (Terralingua undated; Maffi and Oviedo 2000) – supported by WWF and implicit in the other case studies – captures this well (p. 106). This mode of thinking is effectively environmental determinism mediated through a fallacious discourse about culture, community and development. It is ecocentrism with a human face.

Importantly, the advocacy of ecotourism is not simply about balancing two competing priorities, or about trying to combine two separate policies. Rather, the priorities are presented as being *symbiotic* – mutually reinforcing and intrinsically linked. This is not a semantic point. Conservation is presented as being the conservation of local natural capital, which in turn is the favoured basis for development. Development becomes the benefits that accrue from conservation thus defined. Thus, the advocacy of ecotourism effectively ties its conception of development to localised natural limits. In more practical terms, this means that although ecotourism can deliver material benefits, these benefits can only ever be, by their nature, very limited.

This symbiosis between the conservation of natural capital and human development – commonly lauded as a 'holistic' approach – has a chilling effect on how development is conceived of. As Adams points out, conservationists often claim that (objective) scientific certainty underlies their position, and hence place themselves above (subjective) ideology (2001: 69; see also Urry and MacNaghten 1998; Yearley 1996). If the merging of scientific 'fact' relating to the environment is closely linked to formulations of a new, limited development outlook, then political contestation of development is closed off – it becomes closely tied to perceived environmental imperatives. In effect, the mode of development deemed sustainable becomes a Durkheimian 'social fact', an external reality rather than a contested ideology. This study has argued that this is the case, and that the 'symbiosis' argument at the heart of ecotourism's claims to be sustainable development effectively forecloses the sort of thorough debates on rural development that are much needed.

In addition, though it cannot be developed here, it should be noted that facts, their significance and their presentation are contested. Bjorn Lomborg's *The Skeptical Environmentalist: Measuring the Real State of the World* (2001) argues that many of the scientific premises taken as a justification for contemporary conservation initiatives are based on an exaggerated reading of statistics. Further, it has been argued that this, in turn, may be more a product of a society fearful of scientific and technological advance and profoundly disillusioned with development than of any dangers inherent in development itself (Lomborg 2001; Furedi 2002).

Yet even where there are instances of critical natural capital, natural resources that are vital, irreplaceable and without substitutes, why should development for human communities be tied to environmental imperatives? Why not offer communities something better than a life close to nature? There is nothing intrinsically positive in encouraging specific groups of people to remain in a traditional relationship to their land. The thrust of development historically has tended to separate people from a direct dependence on their immediate environment, through urbanisation, trade and the development of division of labour. People in the developed world reap many benefits from this legacy (the ability to travel widely for leisure itself being just one). The talking up of a 'harmonious' relationship between communities and their environs reflects a degraded development agenda that should be challenged rather than lauded as ethical.

While the varied literature on ecotourism emphasises the non-consumptive utilisation of natural resources (e.g. Fennell 2003; UNEP/WTO 2002a: 66 and elsewhere; Goodwin 2000: 97–112; USAID 1996) it should be remembered that the modernisation paradigm inspired conception of development involves transforming the natural world for productive ends. The advocacy of a symbiosis between conservation and development as sustainable development, while sensitised to the destructive aspect of economic development, also embodies a much more static conception of culture than its modernist adversary – the former eschews transformative, thoroughgoing development, typically as 'unsustainable'. It fails to challenge, and in fact celebrates as 'sustainable development', the direct relationship that rural developing world communities have to their natural environment, a relationship that defines their poverty.

Ecotourism as ecodevelopment

Throughout this study it has been implied that 'sustainable development' may not be the most useful way to characterise the symbiosis between conservation and development central to ecotourism. The term is so flexible as to obscure key differences between, for example, those in favour of strong and weak versions respectively. A term that seems more apposite is *ecodevelopment.* Ecodevelopment is a broad term that was used initially to describe attempts to bring environmental and development perspectives together in the 1970s and 1980s, strongly influencing and prefiguring sustainable development (Adams 2001: 114). The emphasis on harmonising development and conservation – or people and nature – in the summaries of the case studies in this chapter and elsewhere in the study is shared by advocates of ecodevelopment. For example, ecodevelopment is described as being premised on 'a new *symbiosis* of man and earth' (Sachs 1979: 113; my italics), and as an 'approach to development aimed at *harmonising* social and economic objectives with ecologically sound management, in a spirit of solidarity with future generations' (ibid.; my italics).

Another definition of ecodevelopment is that it aims 'to pursue economic development that relies for the most part on indigenous human and natural resources and that strives to satisfy the needs of the population, most of all the basic needs of the poor' (Glaeser 1984: 11). Further in this vein, Sachs emphasises 'self reliance', also focusing on the satisfaction of basic needs at the local level (Sachs 1979: 113). The emphasis of both authors, as with the advocates of ecotourism, is that development is conceived of on a *localised* basis, with communities better able to meet their basic needs not on the basis of integration with the world economy, but through local natural resources.

Further, ecodevelopment holds that developing countries should look towards *their own* ecology and culture for development, rather than aspiring to develop in the fashion of the advanced capitalist countries (Potter *et al.* 1999: 69). Hence, ecodevelopment implies that development does not have a universal meaning – it can be conceived differently in different circumstances. Put simply, what is appropriate in one society may be inappropriate in another. Again, ecotourism ICDPs reflect this post-modern, relativist development agenda (Potter *et al.* 1999: 71–4), presenting development as the result of *particular* cultural traditions and natural environments, or, more precisely, the relationship between them.

Overall, ecodevelopment is a response to a sense that development in the past has ridden roughshod over the environment, and has culturally cut people adrift from nature (Adams 2001). It is an explicit appeal to promote a reconnecting of humanity with nature (Hettne 1990). As Urry and MacNaghten point out, such appeals are often associated with 'a wider unease with the modern world' (1998: 16). Such sentiments, they argue, reflect an 'aspiration for more meaningful collective engagement and moral renewal' (ibid.). This outlook is commonplace in the advocacy of ecotourism (Butcher 2003a), which often involves a rejection of mainstream notions of development alongside a deep unease with the experience of the developed world (ibid.).

This kind of ecodevelopment endorses Dasmann's (1976) 'ecosystem people' model, 'ecosystem people' being indigenous rural dwellers who live within one or two ecosystems and whose closer relationship to the land means they are more likely to protect it as the basis of their livelihood. 'Biosphere people', on the other hand, live within a globally linked economic system, characterised by international trade, an extensive division of labour and the use of modern technology in production and distribution – biosphere people do not directly experience the consequences of their development.

The advocacy of ecotourism as sustainable development amounts to advocacy of the ecosystem people model – a localised relationship between people and environment through which the former's direct dependence on the latter encourages a self-sustaining approach. Yet ecosystem people lack the ability to act on nature beyond the local level – ecosystem people cannot intervene in the global environment in a planned, systematic fashion, and hence are subject to the vagaries of climatic change and other natural factors in a way more developed societies (consisting of biosphere people) are not. Neither can ecosystem people aspire to material comforts beyond those that can be achieved on the basis of this relationship. The ecosystem people model that is implicit in the advocacy of ecotourism as development, represents a remarkable wholesale rejection of modern development.

Basic needs and development

The 'development' aspect of ecodevelopment is often associated with poverty relief and the provision of basic needs – it is certainly the case that ecotourism has had a measure of success in promoting the provision of basic needs for rural communities, notwithstanding the opportunity cost of such projects. Yet, as Joseph points out (2001: 148), 'there is a tendency to regard any successful poverty relief programme at the micro level as "local development"', leading to a situation whereby '[t]he concept of local development [. . .] loses any relation to envisaging and working towards other, more holistic and more human, forms of development'. Joseph's view is evident in the advocacy of ecotourism. Here, 'development' effectively refers to small-scale, limited poverty relief. Hence, the advocacy of ecotourism involves a reworking of 'development' away from ongoing human liberation from toil towards an ecocentric harmony with the environment . . . with basic needs for people factored in and talked up as 'sustainable development'.

Conclusion

This chapter has reconstructed the 'symbiosis' argument, the central rationale for ecotourism ICDPs, in the light of the analysis of its constituent parts considered in Chapters 4, 5 and 6. It has noted that symbiosis between conservation and development in theory is reflected in an institutional symbiosis, suggesting that differentiating clearly between 'conservation' and 'development' organisations has become more problematic.

The case studies assert that there is a symbiosis between the tourism industry and conservation, and equate this with sustainable development. Yet, though environmental considerations are obviously important for the tourism industry in general, the bulk of the industry is not organised on the basis of localised natural limits determined by a pre-existing relationship between people and the environment – this can only be said of ecotourism. Elsewhere, tourism development contributes to development through transforming people's culture, their way of life and the way they relate to their natural surroundings. This transformative aspect of development is not intrinsically good, yet neither is it intrinsically bad, as is implied by the favourable comparison of ecotourism ICDPs to other forms of development in the advocacy of the former (Butcher 2003a).

The symbiosis between conservation and development, claimed to be exemplary sustainable development, ties development to strict, localised natural limits. Ecodevelopment, rather than sustainable development, would be a more useful term to use for this mode of development. Symbiosis between conservation and development, viewed in this light, may be less progressive from the perspective of human well-being than its advocates claim. At the very least it should be discussed in terms of a *particular* ideological standpoint on development, that of ecodevelopment, rather than through the (almost) universally accepted rhetoric of sustainable development.

8 Concluding comments

This brief final chapter presents a résumé of the key arguments in the study. It also revisits the neopopulism that characterises the advocacy of ecotourism, in order to situate the conclusions in this wider outlook on development. Finally the chapter offers some suggestions for a prospective research agenda that could develop from the critique offered in this study.

A synopsis of the study

In the introductory chapter, the principal argument for ecotourism as sustainable development in the rural developing world – that it can bring about a symbiosis between conservation and development – was introduced. This was identified in Chapter 3 as a shared emphasis among the case studies and in the more general advocacy too. However, it is also noted in Chapter 1 that sustainable development is, in general, contested. Although very many people argue for sustainable development, there is relatively little agreement on what, in any given instance, it means.

Further on this theme, Chapter 1 also puts forward the view that versions of sustainable development are in a sense socially constructed – logically constituted, but premised upon a number of assumptions which themselves may be open to interpretation. The aim of the study has been to critically examine these assumptions underpinning one version of sustainable development – that explicit in the advocacy of ecotourism.

The first of these assumptions is the neopopulist emphasis on community participation, considered in 'Community participation in the advocacy of ecotourism' (Chapter 4). The general argument from the case studies is that local communities can gain greater control of their destiny through ecotourism-based ICDPs. Neopopulist development is often described as development that communities 'do to themselves' rather than development that is 'done to them', or alternatively as 'endogenous' (Potter *et al.* 1999: 8) or 'bottom up' (ibid.: 69) development. Hence, community participation is a central part of how development is conceived in this formulation. It is typically discussed in terms of 'empowerment' and 'people-centred' development, while related terms such as 'ownership', 'self-sufficiency' and even 'democracy' are invoked, all suggesting a greater level of community control over their destiny.

Chapter 4 also identifies three critical themes. The first of these is that the scope of community participation through ecotourism – the question of what is actually being participated in – is determined prior to participation. Participation is limited to implementation of the projects. The broader issue of political power and influence, which seems beyond the remit of 'empowerment', is hence neglected.

There is also evidence that community participation is viewed as instrumental to conservation – that it provides a channel of influence through which externally decided priorities can be incentivised and operationalised. This questions the view put forward in the case studies, that community participation marks a substantial extension of community control, or empowerment, in development. It also questions the notion of the NGOs featured in the case studies as, relatively speaking, agenda-less bodies whose role is to facilitate the agency of the community.

Further, participation is invariably envisaged as *local*. The relationship between locally conceived development and national development is substantially unexamined by the case studies and, indeed, in the wider literature on ecotourism. This ignores a vital dimension of development as hitherto generally discussed, and potentially restricts the way development is conceived.

The following two chapters consider the character of the development on offer for the communities concerned. 'Tradition in the advocacy of ecotourism' (Chapter 5) argues that the consistent references to traditional culture and traditional knowledge present ecotourism as deferring to the agency of the local community – it is *their* culture, *their* traditions that are invoked. This resonates with the rhetoric of 'empowerment' and 'people-centred development' referred to earlier. It is argued here that this masks a profoundly functional approach to culture, one that tends to conceive of the relationship between rural communities and the environment in a static fashion, in spite of this rhetoric of 'empowerment'. Further, the discourse is strongly influenced by cultural relativism, which succeeds in championing traditional culture while at the same time implicitly denying the possibilities for modern science and technology to play a central role in rural development. For the cultural relativist advocates of 'people-centred' development, the freedom to remain *different* is paramount, but the aspiration for the *same* levels of development and technology as exist in the richer nations is denied . . . in the name of sustainable development.

Also, there is an emphasis on development on the basis of the non-consumption of natural capital, examined in 'Natural capital in the advocacy of ecotourism' (Chapter 6). Indeed, a central assumption of ecotourism is that it can bring development on the basis of conservation, thus resolving the tension between these two aims. What is rarely explicit in either the literature or the case studies is that this constitutes something very close to strong sustainability, a particular version of sustainable development with a strong affinity to ecodevelopment. Yet, throughout the case studies, ecotourism ICDPs are presented simply as 'sustainable development', without qualification or consideration of competing conceptualisations of this strongly normative term.

Chapter 7, 'Symbiosis revisited', shows how these assumptions – community participation, the emphasis on tradition and on the non-use of natural capital – come together in the invocation of a symbiosis between conservation and development in the case studies. Hence, this chapter reconstructs the general rationale for ecotourism in the light of the critique of the specific assumptions put forward in the preceding three chapters. 'Symbiosis' has its own internal logic that rests on these assumptions. By questioning the assumptions, symbiosis is presented in this chapter as a profoundly limiting discourse from the perspective of development.

The development limits of neopopulism evident in ecotourism

The study has critically considered the rationale behind ecotourism ICDPs emanating from a number of case studies. It was argued in Chapter 2 that ecotourism ICDPs emerged from a convergence between strands of thinking in development and in conservation, and that neopopulism is characteristic of this convergence. Ecotourism, it was argued, is a specific example of this general trend. Therefore, the critique of the thinking behind ecotourism can also be considered a limited critique of the wider category of neopopulist development.

Prominent neopopulist Chambers argues forcefully a point that, on the face of it, is self-evident – that the driver behind development policy should be the poverty so many experience in the developing world (Chambers 1983) – a sentiment echoed in all the case studies in this study. However, from this he argues for sustainable rural livelihoods, held to be ways of living that ensure sufficient and secure food and other resources to meet basic needs (ibid.). This approach to tackling poverty is highly influential generally, and also specifically with regard to ecotourism (e.g. Scheyvens 2002). The promotion of sustainable rural livelihoods as a means of engendering development addresses many of the concerns of development's critics – it is small-scale, more personal, less consumptive of natural resources and ultimately, in the eyes of its advocates, potentially far more sustainable. However, historically development has been premised upon expanding scale (companies becoming larger and increasingly global with the consequent scope for establishing economies of scale), and on the development of a division of labour within and between societies (enabling specialisation on the basis of comparative advantage). Also development has involved the creation of cities as centres of commerce and modern living, which have replaced rural living for the vast majority in economically developed societies. As agriculture has become more efficient, labour has been displaced and has provided the manpower for the growth of industries. It is this historical legacy that Chambers, and the advocates of a sustainable rural livelihood, criticise as inappropriate development, or a 'western developmentalism' (McMichael 2001; Hettne 1990).

Of course, Chambers and the neopopulists make many valid points, in that the benefits that have come out of modern development are characterised by their unevenness. Indeed, capitalism as a social system is characterised by combined and uneven development – the fate of the different countries is combined, and we live in a world system from which it is impossible to withdraw, but the benefits gained in some countries are premised upon an exploitative relationship with other parts of the world. But the advocates of community-based sustainable development seem to hold in low regard the technological and material improvements that modern society has brought about. Hence, they do not argue to generalise these benefits, to make them available to all, to make the best and most advanced technology available to help improve health or the latest building techniques to reduce the risk of earthquake or flood damage. Rather, what is advocated is the meeting of 'basic needs', or 'sustainable rural livelihoods', not as a stopgap measure, *but as development itself.* This is the approach of ecotourism ICDPs – to orient rural development around a rural, self-sustaining livelihood that meets basic needs.

Neopopulism has emerged as a key perspective in development thinking (Potter *et al.* 1999: 68; Hettne 1990). Its emergence has brought with it new, innovative thinking, but also the decline of some of the more traditional notions of development. The attempts to combine conservation and development by the NGOs featured in the study, through ecotourism in the developing world, draws heavily on neopopulism, and is presented as innovative and exemplary sustainable development. This perspective is also reflected in the literature on ecotourism and other forms of nature-based tourism in the developing world (e.g. Scheyvens 2002; Goodwin 2000; Honey 1999).

Yet in this context, neopopulism appears to represent not just a change in thinking, but a *retreat* from a discussion of thoroughgoing development. Though offering popular participation locally, the terms of participation in ecotourism ICDPs are shaped around a set of ecodevelopment influenced priorities determined elsewhere. Moreover, neopopulism reflects a diminished view of the potential for widening out these very terms, the terms on which development itself is discussed. It appears to involve a retreat from development at the level of the nation, and an elevation of the importance of distinctly local projects, projects that often eschew national development. Criticisms in the past of 'top down' development and of the 'trickle down effect' of grand projects have merit. However, the NGOs featured in the case studies take a position far in the other direction, elevating the importance of local development and, in the main, failing to link it to broader national strategies.

Also, as has been argued, the language of 'community', which claims to link development and conservation to the culture and the wishes of the community itself, even to the extent of giving the community control or ownership of a project, only does so by *tying culture to nature*; by limiting the agency of the community to the manner in which they can act as nature's guardians. Hence, empowerment is limited to the question of *how* rather than the question of *what* – the latter question, embodying the substantial issue of power and control, is already answered in the philosophy behind ecotourism. This philosophy ties development to the pre-existing relationship between the community and its natural environment, and in citing this as sustainable, eschews development beyond this. Thus, the anthropocentric emphasis on community evident in the advocacy of ecotourism, while it does formally prioritise culture, does so in such a way as to *naturalise* it, tying it to a pre-existing relationship with the natural environment. This is made more vivid if we consider the emphasis on the non-use of natural capital in the case studies.

The moral authority acquired by ecotourism (Butcher 2003a) is based on its ability to combine development and conservation, a premise shared in all the case studies. Yet this premise should be challenged. A starting point would be to logically *separate* conservation and development – to break the neopopulist logic of symbiosis underpinning ecotourism as sustainable rural development – and to conceptualise environmental priorities, and also developmental priorities for people living in rural areas, as *distinct* and *competing*.

This need not preclude community participation. It could, in fact, situate local community participation within the context of national participation and priorities, rather than privilege the local and small scale over the national level. It could also widen the range of possibilities for the communities themselves, who could express preferences beyond how to manage their natural capital as economic resource or for conservation ends.

It could, in theory, benefit the conservation of biodiversity too, where this is deemed to constitute critical natural capital. Most importantly, it could contribute to a rural development agenda that takes seriously the issue of development beyond the most basic needs in the rural developing world.

A research agenda

There are a number of themes arising from the study that could be usefully examined further. Two important ones are listed here.

First, and most importantly, this study has considered the conceptual underpinning of ecotourism, as expressed in the case studies. It has considered the way the concepts employed, and the fashion in which they are employed, constrains discussions on rural development in the developing world. However, further study could look at projects themselves, to examine how, for example, community participation is both facilitated *and constrained* by the funding and conditions on funding emanating from NGOs. In other words, research could examine how these ideas are worked through in practice.

In similar vein, the extent to which traditional knowledge is incentivised by NGOs through funding, alongside the local community's level of support for this perspective, could be examined. However, such research would need to take account of the fact that aid funding is normally conditional upon acceptance of the terms of the funding. Support for aid linked to traditional knowledge is likely to be favoured by local people over no aid at all, yet that does not mean the community's aspirations are reflected in the project aims. Studies of local resident attitudes would need to take account of this wider context in which choices are made and attitudes expressed.

Second, it would be interesting to consider the sense in which sustainable development has become a rhetorical orthodoxy that serves to mask a debate about substantial alternatives. As argued in this study, strong sustainability is presented as sustainable development in the advocacy of ecotourism ICDPs, and this affords the projects a certain moral credibility vis-à-vis those, by implication less sustainable, alternatives.

In many ways, this latter point is an appeal to question the convergence of development and conservation, and to pull them apart, in order to force the issue of prioritisation. This study has implicitly argued for greater priority for development, and less for conservation. Others may argue the opposite. Either way, it is an argument that can only be prompted by accepting that conservation and human development cannot be reconciled in the midst of poverty in the fashion proposed by ecotourism ICDPs.

Notes

2 Ecotourism in development perspective

1 The term 'fortress conservation' is widely used to describe conservation initiatives that neglect the communities living in or adjacent to the areas being conserved (see for example Adams 2001: 270–7). A different phrase coined to refer to the same thing is the 'fences and fines' approach (Wells and Brandon 1992).
2 The term 'human ecology' refers to conceiving of people in the context of the ecosystem, as distinct from ecosystems as *separate* from human communities.
3 The concept of natural capital refers to utilising aspects of the natural world without transforming them. This concept is discussed in Chapter 6.
4 It should be noted that the boundaries between civil society, the state and the commercial sector are blurred – commercial backing and state funding for the NGOs featured in this study suggest that the division, though useful, is limited.
5 These authors use the term 'socio-environmental movement' to emphasise the presentation of the issues as being both environmental *and* social in nature, a view shared by development and conservation oriented NGOs.

3 Pioneers of ecotourism

1 The precautionary principle is widely invoked in conservation circles to emphasise the need to take a precautionary attitude towards development generally. Although the principle is widely accepted, there is much debate as to how it is interpreted. For example, Lomborg, in *The Skeptical Environmentalist: Measuring the Real State of the World* (2001), effectively challenges the application of the principle on the grounds that it holds back important development.
2 Nature Conservancy was in the initial group of NGOs on which information was collected. In fact, as its literature shows, it now takes a stance on ecotourism ICDPs similar to their erstwhile colleagues in Conservation International (e.g. Drumm 2004).
3 The Ecotourism Society was later to rename itself The International Ecotourism Society (TIES). This ecotourism trade body was part of the initial group of NGOs on which data was collected. CI has a close working relationship with TIES.
4 'Meso level' refers to organisations, both governmental and non-governmental, that play a role in linking the local (micro) to the regional or national (macro) in some way.
5 Pro-poor tourism is effectively a type of tourism ICDP, and is discussed in Chapter 6.

6 Natural capital in the advocacy of tourism

1 What, and how big, these benefits are is disputed. While the reference given, written by Pearce and Moran (1994), views these welfare benefits as in imminent danger from development, Lomborg (2001) is far more sanguine about the ability of societies to expand human welfare through economic growth and technological development.

2 While this is a logical argument, the value of biodiversity is itself disputed (Lomborg 2001: Ch. 23).

Bibliography

Interviews with representatives from NGOs

Barnett, Patricia (2000) Director and of Tourism Concern, on 18/4/00, at Tourism Concern's offices in London

Leijzer, Marcel (2002) Private Sector Development Officer (former Tourism Officer) for SNV, on 30/9/02, telephone interview to SNV's offices in the Hague

Sweeting, Jamie (2002) Director, Travel and Leisure Industry Initiatives, The Center for Environmental Leadership in Business (former Tourism Officer) for CI, on 4/9/02, telephone interview to CI's offices in Washington, DC

Woolford, Justin (2002) Tourism Policy Officer, in International Policy Unit of WWF-UK, on 30/5/02, telephone interview to WWF-UK offices in Oxford

Documents (by case study)

CI (Conservation International)

CI (undated a) *CI's Mission*, accessed at www.conservation.org/xp/CIWEB/about accessed on 17/6/03

CI (undated b) *Mediterranean Basin*, accessed at www.biodiversityhotspots.org/xp/Hotspots/mediterranean/?showpage=HumanImpacts accessed on 30/5/03

CI (undated c) *Conservation Programmes: Conservation Enterprise*, accessed at www.conservation.org/xp/CIWEB/programs/conservation_enterprises/cons_enterprise.xml accessed on 24/4/04

CI (undated d) *Ecotourism*, accessed at www.conservation.org/xp/CIWEB/programs/ecotourism/ecotourism.xml, accessed on 30/05/03

CI (undated e) *World Legacy Awards*, accessed at www.ecotour.org/award/index.htm accessed on 15/4/04

CI (undated f) *Hotspots: Hotspots Science*, accessed at www.biodiversityhotspots.org/xp/Hotspots/hotspotsScience/what_are_hotspots.xml accessed on 1/6/04

CI (undated g) *Biodiversity Hotspots, Comparing Hotspots*, accessed at www.biodiversityhotspots.org/xp/Hotspots/hotspots_by_region/ accessed on 1/6/04

CI (undated h) *Conservation Programmes: Ecotourism – Destination Spotlight*, accessed at www.conservation.org/xp/CIWEB/programs/ecotourism/ecotour_dest_spot_gudigwa.xml accessed on 10/04/03

CI (undated i) *Biodiversity Hotspots*, accessed at www.biodiversityhotspots.org/xp/Hotspots/hotspotsScience/ accessed on 20/7/03

CI (undated j) *Travel and Leisure – The Center For Environmental Leadership in Business* accessed at www.celb.org/travel.html accessed on 30/05/03
CI (1999) *Field Reports: Ecotourism*, CI, Washington, DC
CI (2003) *Annual Report*, CI, Washington, DC
CI/Christ C., Hillel O., Matus, S. and Sweeting, J. (2003) *Tourism and Biodiversity: Mapping Tourism's Global Footprint*, UNEP and CI, Washington, DC
CI/Christ, C. (2004) 'A Road Less Travelled', *Conservation Frontlines*, CI, Washington, DC

SNV (Stichting Nederlandse Verijwilligers)

SNV, (undated a) *Who We Are*, accessed at www.snvworld.org/aboutUs/index.cfm?fuseaction=whoWeAre accessed on 26/03/03
SNV (undated b) *History*, accessed at www.snvworld.org/aboutUs/index.cfm?fuseaction=whoWeAre&bb_ID=517 accessed on 14/4/03
SNV (2000) *Strategy Paper*, SNV, The Hague
SNV (2001) *Annual Report*, SNV, The Hague
SNV/Caalders, J. and Cottrell, S. (2001) *SNV and Sustainable Tourism: Background Paper*, SNV, The Hague
SNV/Fisher, S.H. (undated) Tanzanian Cultural Tourism Programme brochure, written, designed and edited by Stephen H. Fisher
SNV/de Jong, A. (1999) *Cultural Tourism in Tanzania; Experiences of a Tourism Development Project*, SNV, The Hague
SNV/Rozenmeijer, N. (2001) *Community Based Tourism in Botswana*, SNV, The Hague
SNV/Schuthof, M. (undated) *Playing a Trump Card: SNV and Local Governance in East and Southern Africa, a Synopsis*, SNV, The Hague

Tourism Concern

Barnett, P. (1995) 'Editorial', *In Focus*, Summer, no. 16, Tourism Concern, London, p. 3
Barnett, P. (1999) 'Editorial', *In Focus*, Summer, no. 33, Tourism Concern, London, p. 3
Botterill, D. (1991) 'A New Social Movement: Tourism Concern, the First Two Years', *Leisure Studies*, no. 10, pp. 203–17
Eber, S. (ed.) (for Tourism Concern and WWF) (1992) *Beyond the Green Horizon: Principles of Sustainable Tourism*, Earthscan, London
Farrow, C. (1995) 'Quepero – Bringing People Together', *In Focus*, Summer, no. 16, Tourism Concern, London, pp. 9–10
Forsythe, T. (for Tourism Concern and WWF) (1996) *Sustainable Tourism: Moving From Theory to Practice*, WWF/Tourism Concern, London
Tourism Concern (undated a) *International Network on Fair Trade in Tourism*, Tourism Concern, London
Tourism Concern (undated b) *1998 to the Present: Tourism and Human Rights*, accessed at www.tourismconcern.org.uk/campaigns/campaigns_human_rights.htm accessed on 13/3/04
Tourism Concern (undated c) *Exploring the World Travellers Code*, accessed at www.tourismconcern.org.uk/what_we_offer/info_for_tourists/travellers_code.htm accessed on 13/3/04

Tourism Concern, (undated d) *2,000: Burma and Lonely Planet* accessed at www.tourismconcern.org.uk/campaigns/campaigns_burma.htm accessed on 10/5/03

Tourism Concern (1995) *In Focus*, (edition titled 'Local Participation: Dream or Reality?'), Summer, no. 16, Tourism Concern, London

Tourism Concern (2000) *Fair Trade in Tourism, Bulletin 2*, Autumn, Tourism Concern, London

Tourism Concern/Mann, M. (2000) *The Community Tourism Guide*, Earthscan, London

Wheat, S. (1994), 'Is There Really an Alternative Tourism?', *In Focus*, Autumn, no. 13, Tourism Concern, London, pp. 2–3

Wheat, S. (1997) 'Editorial', *In Focus*, Spring, no. 23, Tourism Concern, London, p. 3

UNIYE

UNEP/WTO (2002a) *The World Ecotourism Summit Final Report*, UNEP and WTO, Paris

UNEP/WTO (2002b) *The Quebec Declaration on Ecotourism*, UNEP and WTO, Paris (comprises pp. 65–73 of the Final Report)

WWF (World Wide Fund for Nature)

Maffi, L. and Oviedo, G. (2000) *Indigenous and Traditional Peoples of the World and Eco-Region Based Conservation: An Integrated Approach to Conserving the World's Biological and Cultural Diversity*, WWF-International (People and Conservation Unit)/Terralingua

Synergy (for WWF-UK) (2000) *Tourism Certification: An Analysis of Green Globe 21 and other Certification Programmes*, August, WWF-UK/Synergy

Terralingua (undated) *Terralingua*, accessed at www.terralingua.org/ accessed on 13/5/04

WWF-International (undated a) *Linking Tourism and Conservation in the Arctic*, WWF Arctic Tourism Project, Oslo, Norway

WWF-International (undated b) *Conserving Nature, Partnering with People: WWF's Global Work on Protected Area Networks*, WWF-International, Gland, Switzerland

WWF-International (undated c) *WWF Working Locally with Indigenous and Traditional Peoples: Zimbabwe*, accessed at www.panda.org/about_wwf/what_we_do/policy/indigenous_people/on_the_ground/zimbabwe.cfm on 23/4/2004 accessed on 14/2/02

WWF-International (undated d) *Introducing the Concept of Ecoregions and the Global 200*, accessed at www.panda.org/about_wwf/how_we_work/ecoregions.cfm, accessed on 1/6/04

WWF-International (undated e) *List of all Ecoregions*, accessed at www.panda.org/about_wwf/where_we_work/ecoregions/global200/pages/list.htm accessed on 1/6/04

WWF-International (2001a) WWF *Mission and Tourism, Position Statement*, October, WWF-International

WWF-International, (2001b) *Policy Statement: GATS and Responsible Tourism*, September, WWF-International

WWF-International, (2001c) *Tourism Background Paper*, WWF-International

WWF-International (2002) *BP Walks Away from Arctic Power* (news item), dated 28/11/02, accessed at www.wwf.org.uk/news/n_0000000752.asp accessed on 5/11/03

WWF-International/Denman, R. (2001) *Guidelines for Community Based Ecotourism Development*, July, WWF-International

WWF-International/Denton P. (undated) *WWF and HSBC: Delivering Results Where It Matters*, accessed at www.wwf-uk.org/core/about/ta_0000000524.asp accessed on 2/7/04
WWF-Mediterranean Programme (1999) *Responsible Tourism in the Mediterranean: Principles and Codes of Conduct*, WWF-Mediterranean Programme, Rome
WWF-UK (undated a) *What We Do* accessed at wwf.org.uk/core/about/whatwedo.asp accessed on 18/06/03
WWF-UK (undated b) *Research Center: Taking Action*, accessed at www.wwf.org.uk/researcher/programmethemes/tourism/0000000174.asp accessed on 29/08/02
WWF-UK (undated c) *Who We Are*, accessed at www.wwf.org.uk/core/about/who weare.asp accessed on 18/15/03
WWF-UK (undated d) *Chimpanzee Policy*, accessed at www.wwf.org.uk/researcher/issues/rarespecies/0000000088.asp accessed on 15/4/04
WWF-UK (1999) *Ecotourism and Conservation Go Side by Side in Kunene*, accessed at www.wwf.org.uk/news/n_0000000192.asp accessed on 20/5/04
WWF-UK (2000) *Tourism and Carnivores: the Challenge Ahead*, May, a WWF-UK report, WWF-UK, Oxford
See also Eber, S. (ed.) (1992) and Forsythe, T. (1996) joint publications with Tourism Concern, listed under the latter organisation

Other publications

Abrahamsen, R. (2001) *Disciplining Democracy: Development Discourse and Good Governance in Africa*, Zed Books, London
Acott, T., Latrobe, H. and Howard, S. (1998) 'An Evaluation of Deep Ecotourism and Shallow Ecotourism', *Journal of Sustainable Tourism*, vol. 6, no. 3, pp. 238–53
Adams, W.M. (2001) *Green Development: Environment and Sustainability in the Third World, 2nd edition*, Routledge, London
Adams, W.M. and Hulme D. (1998) *Conservation and Communities: Changing Narratives, Policies and Practices in African Conservation*, Community Conservation Discussion Paper, Institute of Development Policy and Management, University of Manchester
Agrawal, A. (1995) 'Dismantling the Divide Between Indigenous and Scientific Knowledge', *Development and Change*, no. 26, pp. 413–39
Akama, J. (1996) 'Western Environmental Values and Nature Based Tourism in Kenya', *Tourism Management*, vol. 17, no. 8, pp. 567–74
Akerman, M. (2003) 'What Does Natural Capital Do? The Role of Metaphor in Economic Understanding of the Environment', *Environmental Values*, vol. 12, pp. 431–48
Allan, W. (1965) *The African Husbandman*, Oliver & Boyd, Edinburgh
Allen, C.M. and Edwards, S.R. (1995) 'The Sustainable Use Debate: Observation from IUCN', *Oryx*, no. 29, pp. 92–8
Amin, S. (1985) 'Apropos the "Green" Movements', in H. Addo, S. Amin, G. Aseniero, A.G. Frank, M. Friberg, F. Frobel, J. Heinrichs, B. Hettne, O. Kreye and H. Seki, *Development as Social Transformation: Reflections of the Global Problematique*, Hodder and Stoughton (for the UN University), Sevenoaks
Arksey, H. and Knight, P. (1999) *Interviewing for Social Scientists*, Sage, London

Aseniero, G., Frank, A.G., Friberg, M., Frobel, F., Heinrichs, J., Hettne, B., Kreye, O. and Seki, H. (eds) (1985) *Development as Social Transformation: Reflections of the Global Problematique*, Hodder & Stoughton (for the UN University), Sevenoaks

Ashley, C., Boyd, C. and Goodwin, H. (2000) 'Pro-Poor Tourism: Putting Poverty at the Heart of the Tourism Agenda', *Natural Resource Perspectives*, no. 51, March, ODI, London

Baker, S. (2005) *Sustainable Development*, Routledge, London

Barrett, C.B. and Arcese, P. (1995) 'Are Integrated Conservation and Development Projects (ICDPs) Sustainable?: On the Conservation of Large Mammals in Sub-Saharan Africa', *World Development*, vol. 23, no. 7

Baskin, J. (1995) 'Local Economic Development: Tourism – Good or Bad?' pp. 103–16, in *Tourism Workshop Proceedings: Small, Medium, Micro Enterprises*, 9–10 March, Land and Agricultural Policy Centre, Johannesburg

Batisse, M. (1982) 'The Biosphere Reserve: a Tool for Environmental Conservation and Management', *Environmental Conservation*, vol. 9, pp. 101–11

Bebbington, A. and Riddell, R.C. (1997) 'Intermediary NGOs and Civil Society Organisations', chapter in D. Hulme and M. Edwards (eds), *NGOs, States and Donors*, Palgrave Macmillan

Beck, U. (1992) *Risk Society: Towards a New Modernity*, Sage, London

Beckerman, W. (1994) 'Sustainable Development: is it a Useful Concept?', *Environmental Values,* vol. 3, pp. 191–209

Beckerman, W. (1995) 'How would you Like Your Sustainability Sir? Weak or Strong? A Reply to My Critics', *Environmental Values*, vol. 4, pp. 169–79

Beckerman, W. (1996), *Through Green Coloured Glasses*, Cato Institute, Cambridge

Berkes F. and Folke, C. (1994) 'Investing in Natural Capital for the Sustainable Use of Natural Capital', pp. 128–49, in A. Jansson , M. Hammer, C. Folke, and R. Constanza (eds), *Investing in Natural Capital: the Ecological Economics Approach to Sustainability*, Island Press, Cambridge

Berle, A.A. (1990) 'Two Faces of Ecotourism', *Audubon*, no. 92

Blaikie, P. (1995) 'Understanding Environmental Issues', pp. 1–30 in S. Morse and M. Stocking (eds), *People and the Environment*, UCL Press, London

Blaikie, P. (2000) 'Development, Post-, Anti-, and Populist: a Critical Review', *Environment and Planning*, vol. 32, pp. 1033–50

Blaikie, P. and Jeanrenaud, S. (1997) 'Biodiversity and Human Welfare', chapter in K.B. Ghimire and M.P. Pimbert (eds), *Social Change and Conservation*, Earthscan, London

Blaxter, L., Hughes C. and Tight, M. (1998) *How To Research*, Open University Press, Buckingham

Boo, E. (1990) *Ecotourism: the Potentials and Pitfalls*, WWF-International, New York

Botterill, D. (1991) 'A New Social Movement: Tourism Concern, the First Two Years', *Leisure Studies*, vol. 10, pp. 203–17

Bottrill, C.G. and Pearce, G. (1995) 'Ecotourism: Towards a Key Elements Approach to Operationalising the Concept', *Journal of Sustainable Tourism*, vol. 3, no. 1, pp. 45–54

Brandt, W. (1980), *North-South: a Programme for Survival*, Pan, London

Brent-Ritchie, J. (1993) 'Crafting a Destination Vision – Putting the Concept of Resident Responsive Tourism into Practice', *Tourism Management*, vol. 14, no. 5, pp. 379–89

Briggs, J. (2005) 'The Use of Indigenous Knowledge in Development: Problems and Challenges', *Progress in Development Studies*, vol. 5, no. 2, pp. 99–114

Briggs, J.P. and Peat, F.D. (1985) *Looking Glass Universe: the Emerging Science of Wholeness*, Fontana, London

Brohman, J. (1996a) 'New Directions for Tourism in the Third World', *Annals of Tourism Research*, vol. 23, no. 1, pp. 48–70

Brohman, J. (1996b) *Popular Development: Rethinking the Theory and Practice of Development*, Blackwell, Oxford,

Bryant, R. and Bailey, S. (1997) *Third World Political Ecology*, Routledge, London

Buckley, P. (1995) 'Critical Natural Capital: Operational Flaws in a Valid Concept', *Ecos: a Review of Conservation*, vol. 16, nos. 3 and 4, pp. 13–18

Budowski, G. (1976) 'Tourism and Conservation: Conflict, Coexistence or Symbiosis', *Environmental Conservation*, no. 3, pp. 27–31

Burgess, A. (1997) *Divided Europe: the New Domination of the East*, Pluto, London

Burns, P. (1999) *An Introduction to Tourism and Anthropology*, Routledge, London

Burr, V. (1995) *An Introduction to Social Constructionism*, Routledge, London

Butcher, J. (2003a) *The Moralisation of Tourism: Sun, Sand . . . and Saving the World?*, Routledge, London

Butcher, J. (2003b) 'Okotourismus – Elend als Urlaubs- Zeil' (translated by Eva Balzar), in *Novo*, May–August edition (also reprinted in *Frietag* magazine)

Butcher, J. (2003c) 'A Bad Package', *Spiked* (online magazine), accessed at www.spiked-online.co.uk archives

Butcher, J. (2006a) 'The United Nations International Year of Ecotourism: a Critical Analysis of Development Implications', in *Progress in Development Studies*, vol. 6, no. 2, pp. 146–56

Butcher , J. (2006b) 'Natural Capital in the Advocacy of Ecotourism as Sustainable Development', *Journal of Sustainable Tourism*, vol. 14, no. 6, pp. 529–44

Butler, R.W. (1990) 'Alternative Tourism, Pious Hope or Trojan Horse', *Journal of Travel Research*, vol. 28, no. 3, pp. 40–5

Butler, R.W. (1992) 'Alternative Tourism: the Thin End of the Wedge', pp. 31–46 in V. Smith and W. Eadington (eds), *Tourism Alternatives, Potentials and Problems in the Development of Tourism*, University of Pennsylvania Press, Philadelphia

Caldwell, L.K. (1984) 'Political Aspects of Ecologically Sustainable Development', *Environmental Conservation*, vol. 11, pp. 299–308

Callinicos, A. (2003) *An Anti-Capitalist Manifesto*, Polity, London

Campbell, A. (1997) *Western Primitivism: African Ethnicity – a Study in Cultural Relations*, Cassell, London

Campfire (undated) *Campfire*, accessed at www.campfire-zimbabwe.org/ accessed on 20/10/03

Carney, D. (ed.) (1998) *Sustainable Rural Livelihoods: What Contribution Can We Make?*, DfID, London

Carr, E.H. (1990) *What is History?*, Penguin, London

Cater, E. (1992) 'Profits from Paradise', *Geographical*, March edition, RGS, London

Cater, E. (1994) 'Ecotourism in the Third World – Problems and Prospects for Sustainability', pp. 69–86 in E. Cater and G. Lowman (eds) *Ecotourism: a Sustainable Option?*, Wiley, Chicester

Cater, E. and Lowman, G. (eds) (1994) *Ecotourism: a Sustainable Option?*, Wiley, London

Cernea, M. (ed.) (for the World Bank) (1985) *Putting People First: Sociological Variables in Rural Development*, Oxford University Press, Oxford

Chambers, R. (1983) *Rural Development: Putting the Last First*, Longman, London

Chambers, R. (1988) 'Sustainable Rural Livelihoods: a Key Strategy for People, Environment and Development', pp. 1–17 in C. Conroy and M. Litvinoff (eds), *The Greening of Aid: Sustainable Livelihoods in Practice*, Earthscan, London

Chambers, R. (1993) *Challenging the Professions: Frontiers for Rural Development*, Intermediate Technology Publications, London
Chambers, R. (1997) *Whose Reality Counts? Putting the First Last*, Intermediate Technology Publications, London
Child, B. (1996) 'The Practice and Principles of Community-Based Wildlife Management in Zimbabwe: the Campfire Programme', *Biodiversity and Conservation*, vol. 5, no. 3, pp. 369–98
Cohen, E. (1972) 'Towards a Sociology of International Tourism', *Social Research*, no. 39, pp. 179–201
Colchester, M. (1997) 'Salvaging Nature: Indigenous Peoples and Protected Areas', chapter in K.B. Ghimire and M.P. Pimbert (eds) *Social Change and Conservation*, Earthscan, London
Cole, S. (1997) 'Anthropologists, Local Communities and Sustainable Tourism Development', chapter in Stabler M. (eds) *Tourism and Sustainability: Principles to Practice*, CABI Publishing, Oxon, pp. 219–30
Connelly, J. and Smith, G. (1999) *Politics and the Environment*, Routledge, London
Constanza, R. (2003) 'Social Goals and the Valuation of Natural Capital', in *Environmental Monitoring and Assessment*, no. 86, pp. 19–28
Cooke, B. and Kothari, U. (2001) 'The Case for Participation as Tyranny', chapter in Cooke, B. and Kothari, U. (eds) *Participation: the New Tyranny?* Zed Books, London, pp. 1–15
Cooke, B. and Kothari, U. (eds) (2001) *Participation: the New Tyranny?* Zed Books, London
Corbridge, S.E. (1993) 'Marxisms, Modernities and Moralities: Development Praxis and the Claims of Different Strangers', *Environment and Planning,* vol. 11, pp. 449–72
Cornwall, A. and Pratt, G. (eds) (2003), *Pathways to Participation: Reflections of PRA*, ITGO Publishing, London
Cotgrove, S. (1982) *Catastrophe or Cornucopia: the Environment, Politics and the Future*, Wiley, Chichester
Cotgrove, S. and Duff, A. (1980) 'Environmentalism, Middle Class Radicalism and Politics', *Sociology Review*, vol. 28, pp. 235–351
Croall, J. (1995) *Preserve or Destroy? Tourism and the Environment*, Calouste Gulbenkian Foundation, London
Crompton, R. (1993) *Class and Stratification*, Polity Press, Oxford
Crotty, M. (1998) *The Foundations of Social Research: Meaning and Perspective in the Research Process*, Sage, London
Crush J.C. (1995) 'Imagining Development', pp. 1–23 in J.C. Crush, (ed.), *Power of Development*, Routledge, London
Dahrendorf, R. (1990) *Reflections on the Revolution in Europe*, Macmillan, London
Dasmann, R. (1976) 'Future Primitive: Ecosystem People Versus Biosphere People', *CoEvolution Quarterly*, no. 11, pp. 26–31
Dickens, C. (1981) *Hard Times*, Penguin Classics, London
DfID (undated) *Tourism Challenge Fund: a DfID Initiative*, accessed at www.challenge funds.org/adfidtcf.htm accessed on 6/7/04
DfID (1999a) *Tourism Challenge Fund,* Business Partnerships Unit, leaflet, DfID, London
DfID (1999b) *Sustainable Livlihood Guidance Sheets* accessed at www.livelihoods.org/info/guidance_sheets_pdfs/section1.pdf accessed on 10/03/06
DfID/DETR (1998) *Sustainable Tourism and Poverty Elimination* (a report on the workshop of 13 October 1998 held by the Department for the Environment, Transport

and Regions and the DfID, in preparation for the UN Commission on Sustainable Development), accessed at www.igc.org/csdngo/tourism/tour_ukgov.htm accessed on 23/05/2002

Doyle, T. and McEachern, D. (1998) *Environmentalism and Politics*, Routledge, London

Drumm, A. (for The Nature Conservancy) (2004) *Ecolodge Guidelines*, The Nature Conservancy, Vancouver

Eber, S. (ed.) (for Tourism Concern and WWF) (1992) *Beyond the Green Horizon: Principles of Sustainable Tourism*, Earthscan, London

Eckersley, R. (1992) *Environmentalism and Political Theory: Towards an Ecocentric Approach*, UCL Press, London

Edwards, M. and Hulme, D. (eds) (1992) *Making a Difference: NGOs and Development in a Changing World*, Earthscan, London

Edwards, M. and Hulme, D. (eds) (1995), *Non-Governmental Organisations: Performance and Accountability: Beyond the Magic Bullet*, Earthscan, London

Ekins, P., Simon, S., Deutsch, L., Folke, C. and De Groot, R. (2003) 'A Framework for the Practical Application of the Concepts of Critical Natural Capital and Strong Sustainability', *Ecological Economics*, no. 44, pp. 165–85

Ellen, R.F. (1986) 'What Black Elk Left Unsaid: On the Illusory Images of Green Primitivism', *Anthropology Today*, vol. 2, no. 6, pp. 8–12

England, R.W. (2000) 'Natural Capital and the Theory of Economic Growth' *Ecological Economics*, no. 34, pp. 425–31

Escobar, A. (1995) *Encountering Development: The Making and Unmaking of the Third World*, Princeton University Press, New Jersey

Farrow, C. (1995) 'Quepero – Bringing People Together', *In Focus*, Summer, no. 16, Tourism Concern, London, pp. 9–10

Featherstone, M. (1991) *Consumer Culture and Postmodernism*, Sage, London

Feifer, M. (1985) *Going Places: the Ways of the Tourist from Imperial Rome to the Present Day,* Macmillan, London

Feldman, S. (1997) 'NGOs and Civil Society: (Un)stated Contradictions', *The Annals of the American Academy of Political and Social Science*, vol. 554, no. 1, pp. 46–65

Fenech, A., Foster, J., Hamilton, K. and Hansell, R. (2003) 'Natural Capital in Ecology and Economics: an Overview', *Environmental Monitoring and Assessment*, no. 86, pp. 3–17

Fennell, D.A. (2003) *Ecotourism: an Introduction*, 2nd edition, Routledge, London

Finch, J. (1986) *Research and Policy: the Uses of Qualitative Methods in Social and Educational Research*, Falmer, Lewes

Flintan, F. (2001) *Women and CBNRM in Namibia. A Case Study of the IRDNC Community Resource Monitor Project (working paper no. 2)*, November, The International Famine Centre, University College Cork, Cork

Forsythe, T. (for Tourism Concern and WWF) (1996) *Sustainable Tourism: Moving From Theory to Practice*, WWF/Tourism Concern, London

Fox, J.A and Brown, L.D. (1998) 'Introduction', pp. 1–47, in J.A. Fox and L.D. Brown (eds), *The Struggle for Accountability: the World Bank, NGOs, and Grassroots Movements*, MIT press, Cambridge, MA

France, L. (ed.) (1997) *The Earthscan Reader in Sustainable Development*, Earthscan, London

Friberg, M. and Hettne, B. (1985) 'The Greening of the World: Towards a Non Deterministic Model of Global Processes', pp. 204–70 in H. Addo, S. Amin,

G. Aseniero, A.G. Frank, M. Friberg, F. Frobel, J. Heinrichs, B. Hettne, O. Kreye and H. Seki, (eds) *Development as Social Transformation: Reflections of the Global Problematique*, Hodder & Stoughton (for the United Nations University), London
Friedman, J. (1992) *Empowerment: the Politics of Alternative Development*, Blackwell, Oxford
Friedman, J. and Weaver, C. (1979) *Territory and Function: the Evolution of Regional Planning*, Edward Arnold, London
Fukuyama, F. (2001) 'Social Capital, Civil Society and Development', *Third World Quarterly*, vol. 22, part 1, pp. 7–20
Furedi, F. (2002) *The Culture of Fear: Risk Taking and the Morality of Low Expectations*, Continuum, London
Galbraith, J.K. (1969) *The Affluent Society*, 2nd edition, Hamish Hamilton, London
Garrod, B. and Fyall, A. (1998) 'Beyond the Rhetoric of Sustainable Tourism?', *Tourism Management*, vol. 19, no. 3, pp. 199–212
Ghai, D. and Vivian, J.M. (1992) 'Introduction', pp. 1–19 in D. Ghai and J.M. Vivian, *Grassroots Environmental Action: People's Participation in Sustainable Development*, Routledge, London
Ghatak, S. (1995) *Development Economics*, HarperCollins, London
Ghimire, K.B. and Pimbert, M.P. (1997) 'Social Change and Conservation: an Overview of Issues and Concepts', Chapter 1 in K.B. Ghimire and M.P. Pimbert (eds), *Social Change and Conservation*, Earthscan, London
Ghimire, K.B. and Pimbert, M.P. (eds) (1997) *Social Change and Conservation*, Earthscan, London
Giddens, A. (1991) *Modernity and Self Identity: Self and Society in the Late Modern Age*, Polity, London
Giddens, A. (1995) *Beyond Left and Right: the Future of Radical Politics*, Stanford University Press, Stanford
Gilbert, V.C. and Christy, E.J. (1981) 'The UNESCO Programme on Man and Biosphere', pp. 701–20 in E.J. Kormondy and J.F. McCormick (eds), *Handbook of Contemporary Developments in World Ecology*, Greenwood Press, Westport, CT
Glaeser, B. (ed.) (1984) *Ecodevelopment: Concepts, Projects, Strategies*, Pergamon Press, Oxford
Glaeser, B. and Vyasulu, V. (1984) 'The Obsolescence of Ecodevelopment?', pp. 23–36, in B. Glaeser (ed.), *Ecodevelopment: Concepts, Projects, Strategies*, Pergamon Press, Oxford
Goeldner, C., McIntosh, R.W. and Brent Ritchie, J.R. (1999) *Tourism: Principles, Practices and Philosophies*, Wiley, London
Goodwin, H. (2000) 'Tourism and Natural Heritage, a Symbiotic Relationship?', pp. 97–112 in M. Robinson *et al.* (eds), *Environmental Management and Pathways to Sustainable Development*, Business Education Publishers, Sunderland
Gorz, A. (1997) *Farewell to the Working Class*, Pluto, London
Gowdy, John M. (1994) 'The Social Context of Natural Capital: the Social Limits to Sustainable Development', *International Journal of Social Economics*, vol. 21, no. 8, pp. 43–55
Graburn, N. (ed.) (1988) *The Anthropology of Tourism: Special Edition of the Annals of Tourism Research*, vol. 10, no. 1
Gunn, C. (1987) 'Environmental Designs and Land Use', pp. 229–47 in J.R.B. Ritchie and C.R. Goeldner (eds), *Travel, Tourism and Hospitality Research: a Handbook for Managers and Researchers*, J. Wiley & Sons, New York

Hall, C.M. (1998) 'Historical Antecedents of Sustainable Development and Ecotourism: New Labels on Old Bottles?' pp. 13–24 in C.M. Hall and A. Lew (eds), *Sustainable Tourism: a Geographical Perspective*, Longman, London

Hall, C.M. and Lew, A. (eds) (1998a) *Sustainable Tourism: a Geographical Perspective*, Longman, London

Hall, C.M, and Lew, A. (1998b) 'The Geography of Sustainable Tourism Development: an Introduction', pp. 1–12 in C.M. Hall and A. Lew (eds), *Sustainable Tourism: a Geographical Perspective*, Longman, London

Hann, C. and Dunn, E. (eds) (1996) *Civil Society: Challenging Western Models*, Routledge, London

Harrison, C., and Burgess, J. (1993) 'The Circulation of Claims in the Cultural Politics of Environmental Change', in A. Hansen (ed.), *The Mass Media and Environmental Issues*, Leicester University Press, Leicester

Harrison, D. and Price, M. (1996) 'Fragile Environments, Fragile Communities? An Introduction', chapter in M. Price (ed.), *People and Tourism in Fragile Environments*, Wiley, Chichester

Hawkins, D. and Khan, M. (1998) 'Ecotourism Opportunities for Developing Countries', pp. 191–201 in W. Theobald (ed.), *Global Tourism*, Butterworth-Heinemann, Oxford

Hays, S.P. (1987) *Beauty, Health and Permanence: Environmental Politics in the United States 1955–1985*, Cambridge University Press, Cambridge

Heartfield, J. (2002) *The Death of the Subject Explained*, Sheffield Hallam University Press, Sheffield

Heinen, J. (1994) 'Emerging, Diverging and Converging Paradigms on Sustainable Development', *International Journal of Sustainable Development and World Ecology*, pp. 22–33

Hertz, N. (2001) *The Silent Takeover*, Heinemann, London

Hettne, B. (1990) *Development Theory and the Three Worlds*, Longman, London

Hettne, B. (1995) *Development Theory and the Three Worlds*, 2nd edition, Longman, London

Hewison, R. (1987) *The Heritage Industry*, Methuen, London

Hilhorst, D. (2003) *The Real World of NGOs: Discourses, Diversity and Development*, Zed Books, London

Hitchcock, M., King, V. and Parnwell, M. (1993) 'Introduction', pp. 1–31, in M. Hitchcock, V. King and M. Parnwell (eds), *Tourism in South East Asia*, Routledge, London

Hobson, J.A. (1965) *Imperialism*, University of Michigan Press, Michigan

Honey, M. (1999) *Ecotourism and Sustainable Development: Who Owns Paradise?*, Island Press, Washington, DC

Honneth, A. (1996) *The Struggle for Recognition: the Moral Grammar of Social Conflicts*, Polity, London (translated by Joel Anderson)

Howarth, D. (2000) *Discourse*, OU Press, Buckingham

Hughes, G. (1995) 'The Cultural Construction of Sustainable Tourism', *Tourism Management*, vol. 16, no. 1, pp. 49–59

Hughes, J. (1990) *The Philosophy of Social Research*, 2nd edition, Longman, Harlow

Hulme, D. and Edwards, M. (1996) *NGOs, States and Donors*, Palgrave Macmillan, London

Inskeep, E. (1991) *Tourism Planning: an Integrated and Sustainable Development Approach*, Van Nostrand Reinhold, New York

IUCN (1991) *Caring for the Earth: a Strategy for Sustainability*, International Union for the Conservation of Nature, Gland

IUCN, UNEP and WWF (1980) *World Conservation Strategy: Living Resource Conservation for Sustainable Development*, International Union for the Conservation of Nature, Gland

Johnston, A. (2005) *Is the Sacred for Sale?: Tourism and Indigenous Peoples*, Earthscan, London

Jordan, A. and Voisey, H. (1998) 'The "Rio Process": the Politics and Substantive Outcomes of "Earth Summit II"', *Global Environmental Change*, vol. 8, pp. 93–7

Joseph, J. (2001) 'Fragmented Dreams', in Eade, D. and Ligteringen, E. (eds), *Debating Development: NGOs and the Future,* Oxfam, Oxford

Klein, N. (2001) *No Logo*, Flamingo, London

Koch, E. (1997) 'Ecotourism and Rural Reconstruction in South Africa', chapter in Ghimire, K.B. and Pimbert, M.P. (1997) *Social Change and Conservation*, Earthscan, London pp. 214–38

Krippendorf, J. (1987) *The Holidaymakers: Understanding the Impact of Leisure and Travel*, Heineman, London

Kumar, K. (1993) 'Civil Society: an Enquiry into the Usefulness of a Historical Term', *British Journal of Sociology*, vol. 44, no. 3

Lane, D. (1984) *Soviet Economy and Society*, Blackwell, London

Leech, K. (2002) 'Enforced Primitivism', pp. 75–94 in T. Jenkins (ed.) *Ethical Tourism: Who Benefits?*, Hodder & Stoughton, London

Lelke, S. (1991) 'Sustainable Development: a Critical Review', *World Development*, vol. 19, no. 6, pp. 607–21

Lenin, V.I. (1996) *Imperialism the Highest Stage of Capitalism*, Junius, London

Lino-Grima, A.P., Horton, S., Kant, S. (2003) 'Introduction: Natural Capital, Poverty and Development', *Environment, Development and Sustainability*, vol. 5, pp. 297–314

Lister, R. (2001a) 'Changing Face of the Arctic', *BBC News Online: World – Americas* dated Wednesday 9th May, accessed at www.news.bbc.co.uk/1/low/world/americas/1318263.stm accessed on 8/11/01

Lister, R. (2001b) 'Clash over Arctic Reserves', *BBC News Online: World – Americas,* dated Wednesday 9th May, accessed at www.news.bbc.co.uk/1/hi/world/americas/1321038.stm accessed on 8/11/01

Logan, B. I. and Moseley, W. G. (2002) 'The Political Ecology of Poverty Alleviation in Zimbabwe's Communal Areas Management Programme for Indigenous Resources (Campfire)', *Geoforum*, vol. 33, no. 1, pp. 1–14

Lomborg, B. (2001) *The Skeptical Environmentalist: Measuring the Real State of the World*, Cambridge University Press, Cambridge

MacCannell, D. (1992) *Empty Meeting Grounds: the Tourist Papers*, Routledge, London

McClaren, D. (1998) *Rethinking Tourism and Ecotravel: the Paving of Paradise and What You Can Do to Stop It*, Kumarian Press, Connecticut

McCormick, J. (1995) *The Global Environmental Movement*, 2nd edition, Wiley, Chichester

McIvor, C. (1997) 'Management of Wildlife, Tourism and Local Communities in Zimbabwe', chapter in K.B. Ghimire and M.P. Pimbert (eds) (1997) *Social Change and Conservation*, Earthscan, London, pp. 239–69

McMichael, P. (2000) *Development and Social Change: a Global Perspective,* 2nd edition, Sage, London

McNeely, J.A. (1984) 'Introduction: Protected Areas are Adapting to New Realities', pp. 1–7 in J.A. McNeely and K.R. Miller (eds), *National Parks, Conservation and Development: the Role of Protected Areas in Sustaining Society*, Smithsonian Institute Press, Washington, DC

McNeely, J.A. (1993) 'Economic Incentives for Conserving Biodiversity: Lessons for Africa', *Ambio*, vol. 22, pp. 144–50

McNeely, J.A. and Miller, K.R. (eds) (1984) *National Parks, Conservation and Development: the Role of Protected Areas in Sustaining Society*, Smithsonian Institute Press, Washington, DC

McShane, T.O. and Wells, M.P. (eds) (2004) *Getting Biodiversity Projects to Work: Towards More Effective Conservation and Development*, Columbia University Press, New York

Maffi, L. and Oviedo, G. (2000) *Indigenous and Traditional Peoples of the World and Eco-Region Based Conservation: an Integrated Approach to Conserving the World's Biological and Cultural Diversity*, WWF-International (People and Conservation Unit)/Terralingua

Marshall, G. (ed.) (1998) *Oxford Dictionary of Sociology*, 2nd edition, Oxford University Press, Oxford

Maunder, D., Myers, D., Wall N. and Miller, R.L. (1995) *Economics Explained*, 3rd edition, Collins Educational, London

May, T. (1993) *Social Research: Issues, Methods and Processes*, Open University Press, Buckingham

Meiksins-Wood, E. (1990) 'The Uses and Abuses of Civil Society', *Socialist Register*, annual edition, Merlin, London

Midgeley, J. (1986) 'Introduction: Social Development, the State and Participation', pp. 1–11 in J. Midgeley, A. Hall, M. Hardiman and D. Narine, *Community Participation, Social Development and the State*, Methuen, New York

Midgeley, J. (2003) 'Social Development: the Intellectual Heritage', *Journal of International Development*, vol. 5, no. 7, pp. 831–41

Milton, K. (1996) *Environmentalism and Cultural Theory*, Routledge, London

Mogelgaard, K. (2003) *Helping People, Saving Biodiversity: an Overview of Integrated Approaches to Conservation and Development*, Occasional Paper, Population Action International, March

Monbiot, G. (2004) *The Age of Consent: Manifesto for a New World Order*, Perennial, London

Morrison, K. (1995) *Marx, Durkheim, Weber: Foundations of Modern Social Thought*, Sage, London

Mort, F. (1989) 'The Writing on the Wall', *New Statesman and Society*, 12 May 1989, pp. 40–1

Mountain Partnership (undated) *Hindu Kush Himalaya in Focus: Indigenous Knowledge*, accessed at www.mountainpartnership.org/initiatives/infocus/hkhimalaya01.html accessed on 18/05/06

Mowforth, M. and Munt I. (1998) *Tourism and Sustainability: New Tourism in the Third World*, Routledge, London

Mulgan, G. (1997) *Life After Politics: New Thinking for the 21st Century*, Demos/HarperCollins, London

Muller, H. (1994) 'The Thorny Path to Sustainable Tourism Development', *Journal of Sustainable Tourism*, vol. 2, no. 1, pp. 131–6

Munt, I. *(1994)* 'Ecotourism or Egotourism?', *Race and Class*, vol. 36, no. 1, pp. 49–60

Murombedzi, J. C. (1999) 'Devolution and Stewardship in Zimbabwe's Campfire Programme', *Journal of International Development*, vol. 11, no. 2, pp. 287–94

Murphy, P. E. (1985) *Tourism: a Community Approach*, Methuen, London

Narman, A. and Simon, D. (eds) (1999) *Development as Theory and Practice: Current Perspectives on Development and Development Cooperation*, Longman, London

Nash, D. (1996) *Anthropology of Tourism*, Pergamon, Oxford

Neale, G. (1998) *The Green Travel Guide*, Earthscan, London

Neumann, R.P. (1997) 'Primitive Ideas: Protected Area Buffer Zones and the Politics of Land in Africa', *Development and Change*, no. 28, pp. 559–82

Oates, J.F. (1999) *Myth and Reality in the Rainforest: How Conservation Strategies are Failing West Africa*, University of California Press, Los Angeles

Offe, C. (1996) *Modernity and the State*, Polity, London

Oppermann, M. (1993) 'Tourism Space in Developing Countries', *Annals of Tourism Research,* vol. 20, pp. 535–56

Osborn, D. and Bigg, T. (1998) *Earth Summit II: Outcomes and Analysis*, Earthscan London

Paehlke, R.C. (1989) *Environmentalism and the Future of Progressive Politics*, Yale University Press, New Haven, CT

Page, S. and Dowling, R. (2002) *Ecotourism*, Prentice Hall, London

Parnwell, M. (1998) 'Tourism, Globalisation and Critical Security in Myanmar and Thailand', *Singapore Journal of Tropical Geography*, vol. 19, no. 2, pp. 212–31

Parsons, T. and Shils, E. (eds) (1954) *Towards a General Theory of Action,* 3rd edition, Harper and Rowe, New York

Pearce, D.W. and Moran, D. (1994) *The Economic Value of Biodiversity*, Earthscan, London

Pearce, F. (1991) 'North-South Rift Bars Path to Summit', *New Scientist*, 22 November, pp. 20–1

Peet, R. and Watts, M. (1996) 'Liberation Ecology: Development, Sustainability and Environment in an Age of Market Triumphalism', pp. 1–45, in R. Peet and M. Watts (eds), *Liberation Ecologies: Environment, Development, Social Movements*, Routledge, London

Pepper, D (1996) *Modern Environmentalism: an Introduction*, Routledge, London

Pieterse, J.N. (1998) 'My Paradigm or Yours? Alternative Development, Post Development, Reflexive Development', *Development and Change*, vol. 29, no. 2, pp. 343–73

Pimbert, M.P. and Pretty, J. (1997) 'Parks, People and Professionals: Putting "Participation" into Protected Area Management' chapter in K.B. Ghimire and M.P. Pimbert (eds) *Social Change and Conservation*, Earthscan, London

Plan Afric (1997) *Appendix to Report for Community Action Project, Report for Ministry of Public Service*, Labour and Social Welfare, Plan Afric, Zimbabwe

Pleumaron, A. (undated) 'Do We Need the International Year of Ecotourism?' Accessed at www.twnside.org.sg/title/iye1.htm accessed on 4/3/2006 (Third World Network web pages)

Pleumaron, A. (1994) 'The Political Economy of Tourism', *The Ecologist*, vol. 24, no. 4, July/August, pp. 142–8

Pleumaron, A. (1995) 'Eco-Tourism or Eco-Terrorism?' *The Environmental Justice Networker*, Winter, no. 6

Poon, A. (1993) *Tourism, Technology and Competitive Strategy*, CABI Publishing, Wallingford

Potter, R., Binns, T., Elliot, J, and Smith, D. (1999) *Geographies of Development*, Longman, London

Powell M. and Seddon, D. (1997) 'NGOs and the Development Industry', *Review of African Political Economy*, no. 71, pp. 3–10

Prentice, R. (1993) 'Community Driven Tourism Planning and Residents' Preferences', *Tourism Management*, vol. 14, no. 3, pp. 218–27

Preston, W. (1996) *Development Theory: an Introduction*, Blackwell, Oxford

Pretty, J. (1994) *Alternative Systems of Enquiry for a Sustainable Agriculture*, IDS Bulletin, no. 25, pp. 37–48

Pretty, J. (1995) 'The Many Interpretations of Participation', *In Focus*, no. 16, Tourism Concern, London, pp. 4–5

Princen, T. and Finger, M. (1994) *Environmental NGOs in World Politics*, Routledge, London

Putnam, R. (2000) *Bowling Alone: the Collapse and Revival of American Community*, Simon & Schuster, New York

Reader, J. (1998) *Africa: a Biography of the Continent*, Penguin, London

Redclift, M. (1990) *Sustainable Development: Exploring the Contradictions*, Routledge, London

Reed, M. (1997) 'Power Relations and Community Based Tourism Planning', *Annals of Tourism Research*, vol. 24, no. 3, pp. 566–91

Reich, C. (1970) *The Greening of America*, Random House, New York

Rich, B. (1994) *Mortgaging the Earth: the World Bank, Impoverishment and the Crisis of Development*, Earthscan, London

Robinson, M. (1993) 'Governance, Democracy and Conditionality: NGOs and the New Policy Agenda', in A. Clayton (ed.), *Governance, Democracy and Conditionality: What Role for NGOs?*, INTRAC, Oxford

Rojek, C. (1995) *Decentring Leisure: Rethinking Leisure Theory*, Sage, London

Rubin, I.I. (1979) *A History of Economic Thought*, Pluto, London

Russell, I. (2000) 'Community Based Tourism: Occasional Studies', *Travel and Tourism Analyst*, no. 5

Sachs, I. (1979) 'Ecodevelopment: a Definition', *Ambio*, vol. 8, nos. 2/3

Scheyvens, R. (1999) 'Ecotourism and the Empowerment of Local Communities', *Tourism Management*, vol. 20, no. 2, pp. 245–9

Scheyvens, R. (2002) *Tourism for Development: Empowering Communities*, Prentice Hall, Harlow

Scott, A. (1990) *Ideology and the New Social Movements*, Unwin, London

Seeland, K. (2000) *Local Knowledge and the Development Process: Cross-Cultural Research on Indigenous Knowledge of Trees and Forests*, paper at Anthropological Studies Association (ASA) conference 2000, at SOAS

Seers, D. (1996) 'The Meaning of Development', *International Development Review*, vol. 11, no. 4, pp. 2–6

Seers, D. (1997) 'The New Meaning of Development', *International Development Review*, vol. 19, no. 3, pp. 2–7

Seligman, A. (1992) *The Idea of Civil Society*, Free Press, New York

Selwyn, T. (ed.) (1996) *The Tourist Image: Myths and Myth Making in Tourism*, Wiley, London

Shah, K. and Gupta, V. (2000) *Tourism, the Poor and Other Stakeholders: Experience in Asia*, ODI and Tourism Concern, London

Sharpley, R. (2000) 'Tourism and Sustainable Development: Exploring the Theoretical Divide', *Journal of Sustainable Tourism*, vol. 8, no. 1, pp. 1–19

Sherman, R. and Webb, R. (eds) (1988) *Qualitative Research in Education: Forms and Methods*, Falmer Press, Lewes

Singh, N.L. and Titi, V. (1995) *Empowerment Towards Sustainable Development*, Zed Books, London

Smith, V. (1989) *Hosts and Guests: Anthropology of Tourism*, Pennsylvania Press, Penn.

Solesbury, W. (2003) *Sustainable Livelihoods: a Case Study of the Evolution of DfID Policy*, ODI working papers, London

Steer, A. and Wade-Gery, W. (1993) 'Sustainable Development: Theory and Practice for a Sustainable Future', *Sustainable Development*, vol. 1, no. 3, pp. 23–35

Stiefel, M. (1994) *A Voice for the Excluded: Popular Participation in Development – Utopia or Necessity?* Zed Books, London

Stohr, W.B. (1981) 'Development From Below: the Bottom Up and Periphery-Inward Development Paradigm', in W.B. Stohr and D.R.F. Taylor (eds), *'Development From Above or Below': the Dialectics of Regional Planning in Developing Countries*, John Wiley, Chichester

Stohr, W.B. and Taylor, D.R.F (eds) (1981) *'Development from Above or Below': the Dialectics of Regional Planning in Developing Countries*, John Wiley, Chichester

Stompka, P. (1992) 'The Year 1989 as Cultural and Civilisational Break', *Communist and Post Communist Studies*, vol. 29, no. 2, pp. 115–29

Strauss, A (1988) *Qualitative Analysis for Social Scientists*, Cambridge University Press, Cambridge

Tacconi, L. (2000) *Biodiversity and Ecological Economics: Participation, Values and Resource Management*, Earthscan, London

Tandon, R. (2001) 'Riding High or Nosediving: Development NGOs in the New Millennium', in D. Eade and E. Ligteringen (eds), *Debating Development: NGOs and the Future*, Oxfam, Oxford

Terralingua (undated) *Terralingua*, accessed at www.terralingua.org/ accessed on 13/5/04

Theopile, K. (1995) 'The Forest as Business: is Ecotourism the Answer?' *Journal of Forestry*, vol. 93, no. 3, pp. 25–7

Therborn, G. (1995) *European Modernity and Beyond: the Trajectory of European Societies 1945–2000*, Sage, London

Tosun, C. (2000) 'Limits to Community Participation in the Tourism Development Process in Developing Countries', *Tourism Management*, vol. 21, pp. 613–33

Touraine, A. (1988) *The Return of the Actor: Social Theory in Post Industrial Society*, University of Minnesota Press, Minnesota

Turner, L. and Ash, J. (1975) *The Golden Hordes: International Tourism and the Pleasure Periphery*, Constable, London

UNDP (United Nations Development Programme) (1993) *Human Development Report*, Oxford University Press, Oxford

UNEP (United Nations Environment Programme) (1993) *The Global Partnership for Environment and Development: a Guide to Agenda 21, Post Rio Edition*, United Nations, New York

Urry, J. (1996) *The Tourist Gaze*, Sage, London

Urry, J. and Macnaughten, P. (1998) *Contested Natures*, Routledge, London

USAID (United States Aid for International Development) (1996) *Win-Win Approaches to Development and the Environment: Ecotourism and Biodiversity Conservation*, Center for Development Information and Evaluation, July, USAID, Washington, DC

Valenzuela, M. (1991) 'Spain: the Phenomenon of Mass Tourism', in A.M. Williams and G. Shaw (eds), *Tourism and Economic Development: Western European Experiences,* 2nd edition, Belhaven, London

Wallace, T. and Lewis, D. (eds) (2000) *New Roles and Relevance: Development NGOs and the Challenge of Change*, Kumarian Press, Connecticut

Warburton, D. (ed.) (1998) *Community and Sustainable Development: Participation in the Future,* Earthscan, London

Watts, M. (2000) 'Development', pp. 166–71 in R. Johnson, D. Gregory, G. Prat and M. Watts (eds), *The Dictionary of Human Geography*, Blackwell, Oxford

WCED (1987) *Our Common Future*, Oxford University Press, Oxford

Wearing, S. and Neil, S. (1999) *Ecotourism: Impacts, Potentials and Possibilities*, London, Butterworth-Heinemann

Weaver, D.B. (1998) *Ecotourism in the Less Developed World*, CABI Publishing, Oxon.

Wells, M. and Brandon, K. (1992) *People and Parks: Linking Protected Area Management*, World Bank, Washington, DC

Wheat, S. (1994) 'Is There Really an Alternative Tourism?', *In Focus*, Autumn, no. 13, Tourism Concern, London

Wheeller, B. (1992) 'Alternative Tourism – A Deceptive Ploy', in C.P. Cooper, and A. Lockwood (eds), *Progress in Tourism, Recreation and Hospitality Management*, Belhaven, London

Wheeller, B. (1993) 'Sustaining the Ego', *Journal of Sustainable Tourism*, vol. 1, no. 2, pp. 121–9

White, S.C. (2000) 'Depoliticising Development: the Uses and Abuses of Participation', in D. Eade (ed.) *Development, NGOs and Civil Society*, Oxfam, Oxford

Williams, E, White, A. and Spenceley, A. (2001) *UCOTA – the Uganda Community Tourism Association: a Comparison with NACOBTA, Pro poor working paper no. 5*, IIED, ODI and CRT, London

Woodwood, C. (1997) 'Report: "Cashing in on the Kruger": the Potential of Ecotourism to Stimulate Real Economic Growth in South Africa', *Journal of Sustainable Tourism*, vol. 5, no. 2

World Bank (1998) *Indigenous Knowledge for Development: a Framework for Action November*, Knowledge and Learning Centre, Africa Region, Washington, DC

Worster, D. (1993) 'The Shaky Ground of Sustainability', pp. 132–45 in W. Sachs (ed.), *Global Ecology: a New Era of Political Conflict*, Zed Books, London

WTO (World Tourism Organisation) (2001) *Tourism 2020 Vision*, WTO, Madrid

WTO (World Tourism Organisation) (2003) *Compendium of Tourism Statistics, 23rd edition*, World Tourism Organisation, Madrid

Yearley, S. (1996) *Sociology, Environmentalism, Globalisation*, Sage, London

Young, Robert C. (1995) *Colonial Desire: Hybridity in Theory, Culture and Race*, Routledge, London

Zadek, S. (2001) *The Civil Corporation*, Earthscan, London

Ziffer, K. (1989) *Ecotourism: the Uneasy Alliance*, CI, Washington, DC

Index